SpringerBriefs in Materials

The SpringerBriefs Series in Materials presents highly relevant, concise monographs on a wide range of topics covering fundamental advances and new applications in the field. Areas of interest include topical information on innovative, structural and functional materials and composites as well as fundamental principles, physical properties, materials theory and design. SpringerBriefs present succinct summaries of cutting-edge research and practical applications across a wide spectrum of fields. Featuring compact volumes of 50 to 125 pages, the series covers a range of content from professional to academic. Typical topics might include

- A timely report of state-of-the art analytical techniques
- A bridge between new research results, as published in journal articles, and a contextual literature review
- A snapshot of a hot or emerging topic
- An in-depth case study or clinical example
- A presentation of core concepts that students must understand in order to make independent contributions

Briefs are characterized by fast, global electronic dissemination, standard publishing contracts, standardized manuscript preparation and formatting guidelines, and expedited production schedules.

More information about this series at https://link.springer.com/bookseries/10111

Nasim Zarrabi

Optoelectronic Properties of Organic Semiconductors

Charge Generation and Recombination
in Next-Generation Photovoltaic Devices

 Springer

Nasim Zarrabi
Department of Physics
Swansea University
Swansea, UK

ISSN 2192-1091   ISSN 2192-1105   (electronic)
SpringerBriefs in Materials
ISBN 978-3-030-93161-2   ISBN 978-3-030-93162-9   (eBook)
https://doi.org/10.1007/978-3-030-93162-9

This Springer imprint is published by the registered company Springer Nature Switzerland AG
The registered company address is: Gewerbestrasse 11, 6330 Cham, Switzerland

To my love, Farid. Thank you for being you!

Preface

From a technological point of view, organic semiconductor-based devices are of significant interest due to their light weight, ease of processability, conformal flexibility and potentially low cost and low embodied energy production. Motivated by these quite unique selling points, the performance of organic semiconductors has been a subject of multi-disciplinary study for more than 60 years with steady progress in applications such as solar cells, transistors, light emitting diodes and various sensors.

The material properties of organic semiconductors are different from inorganic semiconductors. As a result, although the same laws and concepts can be applied to describe the main physical phenomenon, the working principles of organic semiconductor devices are different from their inorganic counterparts. One of the main characteristics that governs the performance of organic semiconductors is their low dielectric constants, meaning they are excitonic at room temperature. A second main feature that dictates the charge carrier recombination and transport properties is the disordered nature of these semiconductors causing low charge carrier mobilities. The work described in this book focuses on these defining elements, and particularly their implications on photovoltaic devices.

The discussion in this book will start with a brief introduction to organic semiconductors in Chap. 1 and a review into the main electro-optical phenomena in organic solar cells in Chap. 2. Essentially, these two chapters aim to provide the readers with enough information and concepts to comprehend what follows. In Chap. 3, a new method is presented for measuring exciton diffusion lengths in organic semiconductors based upon a low-quencher-content device structure. An anomalously large quenching volume is observed that can be assigned to long-range exciton delocalization prior to thermalization. These ultra-low-impurity content organic solar cells are also very useful as model systems to study and engineer trap states. Subsequently, using this approach, it is found that mid-gap trap states are a universal feature in organic semiconductor donor-acceptor blends and unexpectedly contribute to charge generation and recombination. As shown in Chap. 4, this has a profound impact on the thermodynamic limit of organic photovoltaic devices. Having demonstrated this important new insight, in Chap. 5, it is further shown that a definitive link exists between a reduced recombination rate compared to the Langevin rate in some exceptional, high performance material systems and a significant increase in the dissociation rate of charge transfer states upon post-processing of the active layer.

In sum, the work presented in this book delivers important new insight as to the underlying dynamics of exciton generation and diffusion, charge transfer state dissociation, and indeed the ultimate fate of photogenerated free carriers.

Great Britain, Swansea, UK   Nasim Zarrabi
October 2021

Acknowledgements

First, I would like to thank my Ph.D. supervisors Prof. Paul Meredith and Dr. Ardalan Armin for creating a safe, healthy, and scientifically superb research environment for the Sêr SAM group. Paul, thank you for your consistent support, guidance, and positive vibes. It has been an honour for me to be your student. Ardalan, thank you for your enthusiasm toward our various projects, your support, and your continuous encouragement. I cannot possibly acknowledge your positive influence on my career enough.

Additionally, I would like to thank Dr. Oskar Sandberg and Prof. Ivan Kassal for the many fruitful discussions we had during this journey. I sincerely acknowledge all researchers and fellow students at Sêr SAM that accompanied me during my Ph.D., in particular, Dr. Stefan Zesike for his continuous support in the laboratory, Dr. Wei Li for his great contribution to the device fabrication, Drew Riley for his assistance in photophysics measurements, and Nick Burridge and Dr. Greg Burwell for their valuable assistance in proof-reading this book.

To Rhian Jones and Heather Evans, thank you for all the help you have provided me and for all the happy moments you have created. I also extend a special thank to Paul Hughes, Dr. Ryan Bighman and Dr. Dominic Fung for their patient technical assistance in the laboratory.

My deepest gratitude goes to my Mum and Dad, Shahin and Asad, for their unconditional love, support, and encouragement throughout my life. There are no words to describe how grateful I am to have you as my parents. I am also in debt to my one and only brother Naiem for his unconditional support from day one. Elahe, my sister-in-law, you have given me the experience of having a kind and caring sister. I am thankful especially to my amazing friends, Dr. Sahar Basari Esfahani, Dr. Pegah Massoumi and Prof. Safa Shoaee. I admire you all. Thank you for all the quality time, gals!

I would like to express my gratitude to my love, Farid. Thank you for making me laugh every day, and making me feel special, for being the person I can turn to on tough days, and for patiently listening to my complaints.

Finally, I would like to acknowledge the support from Swansea University strategic initiative in Sustainable Advanced Materials (Sêr-SAM) funded by the Sêr

Cymru II Program through the European Regional Development Fund, and Welsh European Funding Office.

Contents

1 Introduction 1
  1.1 Conductivity in Solid State Ordered Semiconductors 1
  1.2 Photoconductivity in Semiconductors 4
  1.3 Organic Semiconductor Materials 5
  1.4 Physics of Organic Semiconductors 7
  1.5 Working Principle of an Organic Solar Cell 9
    1.5.1 Current Voltage Characterisation of a Solar Cell 9
    1.5.2 Photocurrent Generation 10
    1.5.3 Power Conversion Efficiency 13
  1.6 Summary 13
  References 13

2 Electro-optical Phenomena in Organic Solar Cells 15
  2.1 Light Absorption in Molecular Solids 15
    2.1.1 Classical Point of View 15
    2.1.2 Quantum Mechanical Point of View 19
  2.2 Exciton 20
    2.2.1 Generation and Migration 20
    2.2.2 Dissociation and Charge Transfer State Formation 23
  2.3 Charge Transfer State 25
    2.3.1 Dissociation 25
  2.4 Charge Collection 26
    2.4.1 Charge Transport 26
    2.4.2 Bimolecular Recombination 27
    2.4.3 Trap-Assisted Recombination 28
    2.4.4 Surface Recombination 30
  2.5 Theoretical Limit of the Power Conversion Efficiency of a Solar
    Cell 30
    2.5.1 Charge Transfer State as the Effective Energy Gap
       of an Organic Solar Cells 32
  2.6 Aim and Structure of the Book 34
  References 34

3   Anomalous Exciton Quenching in Organic Semiconductors 37
    3.1   Introduction 37
    3.2   Theoretical Framework 38
    3.3   Experimental Results 41
    3.4   Conclusion 45
    References 46

**4   On the Effect of Mid-Gap Trap States on the Thermodynamic
    Limit of OPV Devices** 49
    4.1   Introduction 49
    4.2   Ultra-Sensitive EQE_{PV} Measurements and the Failure
          of Reciprocity 51
    4.3   Origin of the Low-Energy Sub-Gap Absorption Features
          in the EQE_{PV} 54
    4.4   Charge Generation and Recombination via Mid-Gap States 61
          4.4.1   Modified Shockley Read Hall (SRH) Theory 64
    4.5   The Two-Diode Model and the Origin of the Ideality Factor
          in OSCs 67
    4.6   Impact on the Detectivity of Organic Photodetectors 72
    4.7   Conclusion 74
    References 75

**5   Relating Charge Transfer State Kinetics and Strongly Reduced
    Bimolecular Recombination in Organic Solar Cells** 79
    5.1   Introduction 79
    5.2   Device Characterisation and Transport Measurements 81
    5.3   Theoretical Framework 85
    5.4   Experimental Result and Discussion 87
    5.5   Conclusion 91
    References 92

6   Outlook 95

Appendix: Extra Experimental Methods 99

Acronyms

a	Quencher Radius
$\boldsymbol{a}$	Thermalisation Length
α	Absorption Coefficient
A	Acceptor
A	Absorbance
BHJ	Bulk Heterojunction
β_{Bulk}	Bulk Recombination Rate Constant
β_{en}	Encounter Rate Constant
β_{L}	Langevin Recombination Rate Constant
β_{SRH}	Trap-Assisted Recombination Rate Constant
c	Speed of Light
$\boldsymbol{c}$	Quencher Concentration
CT	Charge Transfer
D	Diffusion Constant
D	Donor
DOS	Density of States
E	Electric Fields
EA	Electron affinity
EQE_{PV}	Photovoltaic External Quantum Efficiency
EQE_{LED}	Electroluminescence Quantum Efficiency
ε	Energy
ε_{V}	The Upper Edge of the Valence Band
ε_{C}	The Lower Edge of the Conduction Band
ε_{g}	Bandgap Energy
ε_{F}	Fermi Energy
ε_{Fn}	Electron Quasi-Fermi
ε_{Fp}	Hole Quasi-Fermi
ε_{B}	Binding Energy
ε_{r}	Dielectric Constant
ε_{0}	Vacuum Permittivity
η_{Abs}	Photon Absorption Efficiency

η_{Ex}	Exciton Dissociation Efficiency
η_{CT}	Charge Transfer State Dissociation Efficiency
η_{Coll}	Charge Collection Efficiency
FF	Fill Factor
$g(\varepsilon)$	Density of states
γ_{Geo}	Geometrical Reduction Factor
γ	Reduction Factor
H	Enthalpy
H'	Perturbation Hamiltonian
HOMO	Highest Occupied Molecular Orbital
$\hbar$	Planck constant $= 6.582 \times 10^{-16}$ eV.s
IP	Ionization Potentials
IQE	Internal Quantum Efficiency
J	Current
J_{Ph}	Photo Current
J_D	Dark Current
J_0	Dark Saturation Current
$\mathbf{k}$	Extinction coefficient
$\bar{\mathbf{k}}$	Wave Vector
k_B	Boltzmann Constant $\approx 8.61 \times 10^{-5}$eV/K
k_d	Dissociation Rate Constant
k_f	Geminate Recombination Rate constant
L_{diff}	Diffusion Length
LUMO	Lowest Occupied Molecular Orbital
λ	Wavelength
λ_{CT}	Charge Transfer Reorganisation Energy
λ_t	Charge Trap Reorganisation Energy
μ_n	Electron Mobility
μ_p	Hole Mobility
$\bar{\mu}$	Dipole Operator
n	Electron Density
n_{id}	Ideality Factor
$\tilde{\mathbf{n}}$	Complex Refractive Index
$\mathbf{n}$	Refractive Index
N_C	Effective Density of State in the Conduction Band
N_V	Effective Density of State in the Valence Band
OLED	Organic Light Emitting Diode
OPV	Organic Photovoltaics
p	Hole Density
PCE	Power Conversion Efficiency
PL	Photoluminescence
PV	Photovoltaics
ϕ_{Sun}	Spectral Flux Density of the Sun
ϕ_{BB}	Spectral Flux Density of a Black Body
ψ_{el}	Electronic State

ψ_{vib}	Vibrational State
q	Elementary Charge
R	Recombination Rate
R_S	Series Resistance
R_{Sh}	Shunt Resistance
r_C	Coulomb Capture Radius
S	Entropy
SNR	Signal to Noise Ratio
σ	Absorption Cross Section
σ_c	Conductivity
σ_d	Disorder Parameter
T	Absolute Temperature
TAS	Transient Absorption Spectroscopy
τ_D	Exciton Lifetime
V	Voltage
V_{OC}	Open Circuit Voltage
V_{OC}^{Rad}	Radiative Limit of the Open Circuit Voltage
QY	Quenching Yield

Chapter 1
Introduction

Abstract The physical characteristics of a substance, such as electrical and optical properties, are determined by the material components and their atomic and molecular arrangement. In the world of photovoltaic devices, the material properties of semiconductors such as charge carrier mobilities and the bandgap are of particular interest. In this chapter, some of the fundamental concepts of semiconductors relevant to the photovoltaic effect were reviewed. The optoelectronic properties of organic semiconductors were explored, and finally, the basic working principles of an organic solar cell are discussed.

1.1 Conductivity in Solid State Ordered Semiconductors

According to the measurable macroscopic property of electric conductivity σ_c, which is the ability of a material to conduct electric current, solid materials are, roughly, classified to three groups: Conductors, often known as metals with the conductivity in the order of 10^7 (S/m), insulators with the conductivity in the order of 10^{-17}(S/m) and semiconductors with the conductivity sitting between the two extremes. Conductivity is proportional to the density n of free carriers (being electrons or holes) and the mobility of free carriers μ in a material system

$$\sigma_c = nq\mu \tag{1.1}$$

where q is the elementary charge [1]. Like any other macroscopic property, conductivity can be inferred from microscopic models.

Let's take lithium (Li), the first metal in the periodic table, as an example. Li has the electron configuration of $1s^2 2s^1$. When Li atoms come together, the electron in the 2s orbital of one Li atom shares space with a corresponding electron in the neighbouring atoms to form molecular orbitals in the same way that a covalent bond is formed. In the most stable form of Li crystal, the body centred cubic structure, each Li atom is surrounded by eight neighbouring Li atoms organised into a cubic array, so that the sharing occurs between each of the atoms with all 8 neighbouring atoms and each of those eight in turn is being in touch with eight other atoms and

N. Zarrabi, *Optoelectronic Properties of Organic Semiconductors*,
SpringerBriefs in Materials,
https://doi.org/10.1007/978-3-030-93162-9_1

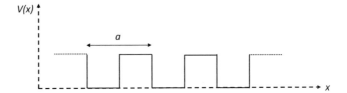

Fig. 1.1 Schematic of periodic potential $V(x)$

so on, to form the whole substance [2]. Since all the electrons in the outer shell of the Li atoms are identical, they cannot occupy the same molecular orbital (due to the Pauli exclusion principle) [3]. As a result, atomic orbitals merge, providing a great number of available molecular orbitals for electrons to occupy over which the electrons are delocalized in the whole structure. The free electron model, although oversimplified, has often been used to represent the electronic structure of metals. In this model the interaction between the valence electron and the lattice ions and the interaction of the valence electrons with each other are neglected.

In a more sophisticated model for perfectly ordered solid materials the potential of the lattice ions needs to be considered. The Kronig–Penny model in one dimension is a simplified model for an electron in a periodic lattice. The possible energy states that an electron can have is determined by solving the time independent Schrodinger equation

$$\left(-\frac{\hbar^2}{2m}\frac{d^2}{dx^2} + V(x) \right)\psi(x) = \epsilon\psi(x), \tag{1.2}$$

for an infinite periodic potential $V(x)$ with period of a (See Fig. 1.1). The solution to the Schrodinger Equation which is the wave-function of an electron must be in form of

$$\psi(x) = exp(i\bar{k}x)u(x), \tag{1.3}$$

known as Bloch function, where $\bar{k}$ is the wave vector and $u(x + a) = u(x)$ is a periodic function of x [4]. The resulting energy eigenvalues $\epsilon_n(\bar{k})$ are a continuous function of the wave vector denoted with an index $n \in \mathbb{Z}^+$. The set of energies for an index n and for all possible $\bar{k}$ is called an energy band. The range of energies between bands where there is no solution for $\epsilon_n(\bar{k})$ is called a bandgap [5]. The energy band above and bellow the bandgap are called the conduction band and the valence band, respectively [6]. In Fig. 1.2 a simplified schematic of an energy band diagram is shown. Here ϵ_V is the upper edge of the valence band and ϵ_C is the lower edge of the conduction band.

In this picture the occupation of the energy states which is also directly related to the width of the bandgap ϵ_g defines the electronic properties of crystalline solids. The probability distribution of the number of electrons of energy ϵ is described by the Fermi–Dirac function

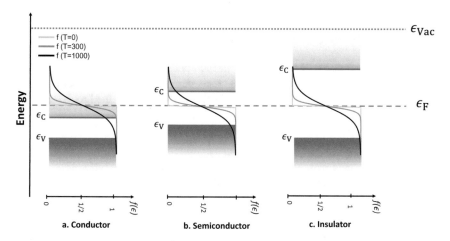

Fig. 1.2 Schematic of energy band diagram and the Fermi distribution function at different temperature for **a** a conductor, **b** a semiconductor and, **c** an insulator

$$f(\epsilon) = \frac{1}{\exp(\frac{\epsilon - \epsilon_F}{k_B T}) + 1},$$ (1.4)

where ϵ_F (Fermi energy) is a hypothetical energy level with probability $\frac{1}{2}$ of being occupied at any temperature T and k_B is the Boltzmann constant [7].

In a conductor (metal), the Fermi level sits in the conduction band Fig. 1.2a. As a result at $T = 0°K$ the conduction band is only partially full. So the electron can move freely and conduct electrical current. In an insulator the Fermi level sits in the middle of the gap and the bandgap is large Fig. 1.2c. Even at room temperatures ($T \approx 300°K$) the valence band is completely full and the conduction band is empty so there is no room for electrons to move. As a result the conductivity is infinitesimal. In an intrinsic crystalline inorganic semiconductor, gallium arsenide (GaAs) as an example, the Fermi level sits within the gap and the bandgap is relatively small Fig. 1.2b. At room temperature, there is a small probability for an electron to be found in the conduction band. However, for a high enough temperature (dependent on the bandgap of the semiconductor) electrons can have enough energy to be promoted to conduction band and leave an empty energy state behind in the valence band (hole). The electron in the conduction band is free from the lattice (ions) potential so that the material can conduct electrical current with rather high mobility ($\sim 10^3$ cm^2/Vs)

In thermal equilibrium the density of electrons (n) and holes (p) in the conduction band and the valence band of a semiconductor are given by

$$n = \int_{-\infty}^{+\infty} g(\epsilon) f(\epsilon) d\epsilon = N_C \exp\left(-\frac{\epsilon_C - \epsilon_F}{k_B T}\right),$$ (1.5)

$$p = \int_{-\infty}^{+\infty} g(\epsilon)[1 - f(\epsilon)]d\epsilon = N_V \exp\left(-\frac{\epsilon_F - \epsilon_V}{k_B T}\right), \tag{1.6}$$

here $g(\epsilon)$ is the density of state (DOS) in the conduction (valence) band which is defined as the number of electron (hole) states per volume per energy interval. For the conduction band in a three dimensional (3D) semiconductor the DOS is given by

$$g(\epsilon) = 4\pi \left(\frac{2m_e^*}{h^2}\right)^{\frac{3}{2}} (\epsilon - \epsilon_C)^{\frac{1}{2}}, \tag{1.7}$$

(a similar expression can be derived for valence band) and N_C (N_V) is the effective DOS (number states accessible within $k_B T$) of the conduction (valence) band which is defined as $N_{C(V)} = 2\left(\frac{2\pi m_{e(h)}^* k_B T}{h^2}\right)^{\frac{3}{2}}$ and $m_{e(h)}^*$ is the effective mass of electron (hole). The detail of the derivations can be found in Wurfel's textbook [5]. For an intrinsic semiconductor in thermal equilibrium the electrons in the conduction band are originated from the valence band and

$$np = n_i^2 = N_C N_V \exp\left(\frac{-\epsilon_g}{k_B T}\right). \tag{1.8}$$

Equation (1.8) is know as mass action law, which states that $n = p = n_i$.

1.2   Photoconductivity in Semiconductors

One of the most interesting properties of semiconductor materials which gives rise to many device applications is photoconductivity. In this phenomenon, by absorbing a photon with energy higher or just equivalent to the bandgap (excitation) electrons are promoted to an energy state in the conduction band. As the result, the semiconductor's conductivity increases with light absorption. This makes them applicable to be used in photovoltaic (PV) devices such as photodetectors and solar cells [8].

Following light absorption, electrons (holes) thermalise to the lowest available energy state within the conduction (valence band) in a very short time scale ($\sim 10^{-12}$ s). At this stage the system is in an inter-band (local) equilibrium state also known as quasi-equilibrium. The local distribution of the electrons (holes) can be described by $f_C(\epsilon)(f_V(\epsilon))$ shown with blue (red) distribution function in Fig. 1.3 and the local density of electrons and holes are given by [5]:

$$n = N_C \exp\left(-\frac{\epsilon_C - \epsilon_{Fn}}{k_B T}\right), \tag{1.9}$$

$$p = N_V \exp\left(-\frac{\epsilon_{Fp} - \epsilon_V}{k_B T}\right). \tag{1.10}$$

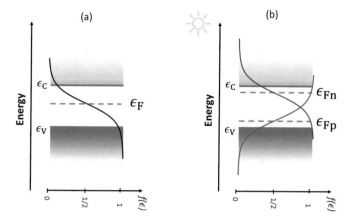

Fig. 1.3 Schematic of energy band diagram and the Fermi distribution function **a** in the dark (in equilibrium condition) **b** under illumination (in quasi-equilibrium condition)

where ϵ_{Fn} and ϵ_{Fp} are the electron and hole quasi-Fermi level, respectively.

In this quasi-equilibrium condition both electron and hole densities are greater than n_i. The mass action law can be written in a more general form as:

$$np = N_C N_V \exp\left(\frac{-\epsilon_g}{k_B T}\right) \exp\left(\frac{\epsilon_{Fn} - \epsilon_{Fp}}{k_B T}\right) = n_i^2 \exp\left(\frac{\epsilon_{Fn} - \epsilon_{Fp}}{k_B T}\right). \qquad (1.11)$$

1.3 Organic Semiconductor Materials

Organic semiconductors are solids predominantly made up from carbon (C) and hydrogen (H) and at times heteroatoms such as oxygen (O), nitrogen (N) and sulfur (S), fluorine (F) and chlorine (Cl). They can be in form of molecular crystalline, amorphous molecular and polymers. Fig. 1.4a to c shows one example of each form.

Independent of the solid state form, the semiconductor properties of these materials are directly derived from the chemical properties of carbon. Atomic carbon has 6 electrons and in its ground state they have an electron configuration of $1s^2 2s^2 2p^2$. When other atoms approach to form a molecule the atomic orbitals in the outer shell hybridize and form hybridized molecular orbitals. One of the electron configurations which is of interest is sp^2 hybrid orbital. Consider the ethene (C_2H_4) molecule shown in Fig. 1.5a which is the simplest molecule with sp^2 hybrid orbitals. Each carbon atom has three $2sp^2$ orbitals in an x-y plane and one $2pz$ in the z direction. The three $2ps^2$ orbitals generate three σ-bonds, two with hydrogen atoms and one with the other carbon atom. The two electrons in $2pz$ orbitals of each atoms will be paired and generate a π-bond. Thus the two carbon atoms in ethene are bonded together with one σ-bond with the orbital probability density centred around the axis

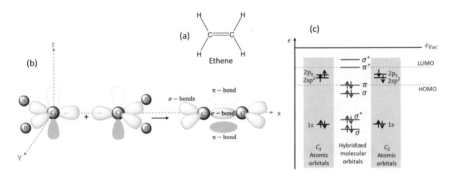

(a). Pentacene

(c). PCDTBT

(b). ITIC

Fig. 1.4 Molecule structure of **a** a crystalline, **b** a small molecule organic, and **c** a polymeric semiconductor

Fig. 1.5 **a** Molecular structure of ethene **b** Three $2sp^2$ hybrid orbitals in the x-y plane and one $2pz$ orbital orthogonal to the plane shown for each carbon atom. The resulting σ- and π- bonds are shown for the ethene molecule. **c** Simple energy level diagram illustrating the formation of HOMO and LUMO energy level and the energy gap

joining two atoms and one weak π-bond with the orbital probability density above and bellow the axis which is delocalized over the whole molecule (see Fig. 1.5b). By applying the linear combination of atomic orbitals (LCAOs) technique the energies of the molecular orbitals can be calculated. A simple energy diagram together with the electron configuration of ethene is shown in Fig. 1.5c [9].

The highest occupied molecular orbital (HOMO) is the bonding π-orbital and the lowest unoccupied molecular orbital (LUMO) is the anti-bonding π^*-orbital. In an organic semiconductor the energy difference between the LUMO and the HOMO is called the energy gap and analogous to the bandgap (in inorganic semiconductors) it dictates the semiconducting properties [10, 11].

1.4 Physics of Organic Semiconductors

From a technological point of view, organic semiconductors are of particular interest. They have typical properties of plastics being lightweight and flexible [12–14]. They are usually soluble in organic solvent so that they can be solution processed by employing simple techniques like spin-coating and ink-jet printing. Moreover, the photophysical proprieties of organic semiconductors can be modified by chemical synthesis [15–18]. Thus, specific materials can be tailored for certain device applications from light emitting diodes to solar cells, photodetectors and optical sensors.

The field of organic semiconductor devices was established long after their inorganic counterparts thus, solid-state semiconductor device physics has been conceptualized on the working principles of inorganic devices. As a result, many of the terminologies in the field of organic semiconductors have been borrowed from inorganic device physics. However, one should keep in mind that although the same laws of physics are applied to both groups of devices, the working principles originating from the material properties, are quite different [19].

The first main difference between organic and inorganic semiconductors is related to the ability of the material to polarize when subjected to electric field which dictates the ease of charge movement in the material. The microscopic measure for this ability is the dielectric constant ε_r. In inorganic semiconductors $\varepsilon_r \approx 12$ and in organic semiconductors $\varepsilon_r \approx 3\text{--}5$ [20, 21].

The second main difference between organic and inorganic semiconductors is related to the binding of the constitute elements in these materials that was discussed earlier. In inorganic semiconductors atoms are strongly bound together via covalent bonds which results in bands in the electronic structure (conduction band and valence band). However, molecules in organic semiconductors, are bound together via weak Van der Waals forces. The weak coupling of the molecules results in individual molecular orbitals (or in the case of crystalline organic semiconductors very narrow valence and conduction bands). The energies vary randomly between the sites (molecules or different segment of a polymer) as a result of different environmental fluctuations that each molecule experiences. The density of energy states for a randomly distributed assembly can be described by the Gaussian distribution function with the mean value of ϵ_0

$$g(\epsilon) = \frac{1}{\sqrt{2\pi}\,\sigma_d} \exp\left(\frac{(\epsilon - \epsilon_0)^2}{2\sigma_d^2}\right), \tag{1.12}$$

where σ_d which is the standard deviation of the distribution known as the disorder parameter (see Fig. 1.6) [22].

In an inorganic semiconductor when an electron is promoted to the conduction band (and leaves a hole in the valence band), by thermal activation or optical excitation, it will straight away become free from the Coulombic attraction ($F_{\text{Coulomb}} = \frac{q^2}{4\varepsilon_r\varepsilon_0 r^2}$) of the lattice ions due to the large value of ε_r. Furthermore, the electron wave function (Bloch function) is delocalized over the whole conduction

Fig. 1.6 Schematic of
density of states $\rho(\epsilon)$ in
organic semiconductors

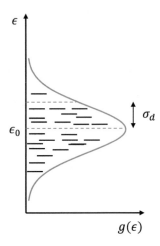

band meaning that the electron is free to move with rather high mobility (10^3–$10^4\,\mathrm{cm^2V^{-1}s^{-1}}$).

In contrast, optical excitation of an organic semiconductor creates a localized excited molecule (ε_r is small so is the F_{Coulomb}). The electron and hole pair, so-called 'exciton', has a very short lifetime (<1 ns) during which the excitation energy may transfer from one molecule to the neighboring molecule (exiton diffusion). The creation of free charge carriers, in this picture, requires an extra driving force to overcome the exciton binding energy [23, 24]. To facilitate exciton dissociation a second material with higher electron affinity, a so called 'electron acceptor', is introduced to the first material matrix, so called 'electron donor'. Exciton dissociation happens via an intermediate state known as charge transfer (CT) state at the donor-acceptor interface (the details will be further discussed in the next chapter). Moreover, migration of free charges, created upon exciton dissociation, occurs via an incoherent transfer mechanism, known as hopping, in the DOS. As a result, the mobility of charge carriers in organic semiconductor is relatively low (10^{-5}–$10^{-3}\,\mathrm{cm^2V^{-1}s^{-1}}$).

The above-mentioned properties play an important role in the operational principles of the devices made from organic semiconductors. Deep understanding of the fundamental processes based on these properties is crucial both from material synthesis and device structure design perspectives.

This book presents a handful of recent findings of organic semiconductors, which are mainly applicable to organic photovoltaics (OPV) and, more specifically, organic solar cells unless stated otherwise. Hence, in the next section the basic concepts and working principles of an organic solar cell will be presented.

1.5 Working Principle of an Organic Solar Cell

A solar cell is a device that converts light energy to electric energy *via* the photovoltaic effect. In this process, a semiconductor material, also known as the active layer, is used to absorb light energy and convert it to voltage and electric current. In efficient organic solar cells, the active layer is made of an interpenetrating network, known as bulk hetrojuction (BHJ), of two organic semiconductors [25, 26]. The two organic semiconductors are known as the donor and the acceptor. The donor is usually a π-conjugated polymer and it is responsible for most of the light absorption and the acceptor is a small molecule. Conventionally, small molecule fullerenes were used extensively as the acceptor, but most recently the new generation of non-fullerene small molecules have shown outstanding performance [27, 28]. In Fig. 1.7 some example of organic donor and acceptor material systems are shown.

In a device structure the active layer is sandwiched between two electrodes. To allow light into the device one electrode must be transparent. For this purpose usually an indium tin oxide (ITO) coated glass substrate is used. For the top electrode a layer of a metal such as aluminum (Al) or silver (Ag) is used. In order to facilitate charge collection at the electrodes, an interlayer is used to adjust the work function[1] of the ITO and the top electrode (see Fig. 1.8).

1.5.1 Current Voltage Characterisation of a Solar Cell

As an electronic component an organic solar cell is analogous to a thin film diode. The current voltage (J-V) characteristic of a solar cell is typically given by

$$J(V) = J_D(V) + J_{Shunt}(V) - J_{Ph} = J_0 \left[\exp\left(\frac{q[V - JR_S]}{n_{id}k_B T} \right) - 1 \right] + \frac{V - JR_s}{R_{Sh}} - J_{Ph}, \quad (1.13)$$

where J_0 is the dark saturation or recombination current density, J_{Ph} is the photocurrent density generated by the the solar cell when it is exposed to a light source, R_S is the series resistance of the external circuit including the sheet resistance of the electrodes, R_{Sh} is the shunt resistance corresponding to the non-ideal leakage current density caused by defects in the active layer, n_{id} is the diode ideality factor which describes how closely the diode performs compared to an ideal Shockley type diode ($n_{id} = 1$). n_{id} is determined by the dominant recombination process in the diode. It has been shown that for organic solar cells n_{id} has a value between one and two [8]. In Chap. 4, this matter will be discussed, extensively. A schematic picture of the equivalent circuit of an organic solar cell is shown in Fig. 1.9

---

[1] Electronic work function is a measure of the amount of energy (or work) required to withdraw an electron from a metal surface.

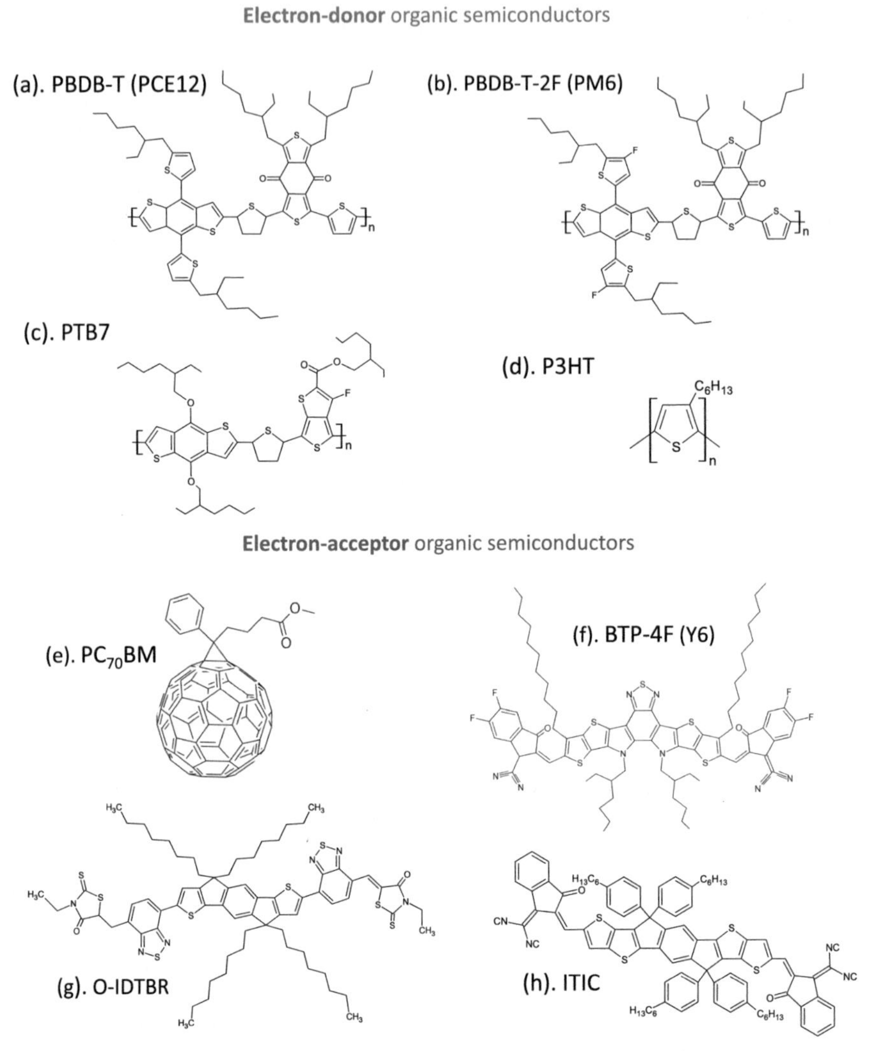

Electron-donor organic semiconductors

(a). PBDB-T (PCE12)

(b). PBDB-T-2F (PM6)

(c). PTB7

(d). P3HT

Electron-acceptor organic semiconductors

(e). PC$_{70}$BM

(f). BTP-4F (Y6)

(g). O-IDTBR

(h). ITIC

Fig. 1.7 Examples of electron donor: **a** PBDB-T (**PCE12**), **b** PBDB-2F (**PM6**) **c** PTB7, **d** P3HT and electron acceptor: **e** PC$_{70}$BM, **f** BTP-4F (**Y6**), **g** O-IDTBR, **h** ITIC organic semiconductors. See Appendix for full chemical names

1.5.2 Photocurrent Generation

The photocurrent generation in organic solar cells can be explained in sequential steps as follows (see Fig. 1.10)

i: Photon absorption and exciton formation

ii: Exciton diffusion and dissociation to charge transfer state

Fig. 1.8 A schematic picture of a BHJ organic photovoltaic device

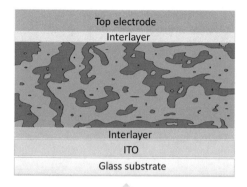

Fig. 1.9 A schematic picture of an equivalent circuit of a thin film solar cell

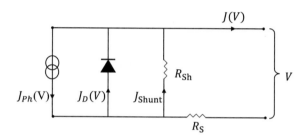

iii: Charge transfer state dissociation

iv: Charge collection.

The details of each step will be discussed in Chap. 2 [20]. The product of the efficiencies of these sequential steps defines the external quantum efficiency EQE_{PV} of a device which is the number of electrons that reach the external circuit per incident photon

$$EQE_{PV} = \eta_{Abs} \times \eta_{Ex} \times \eta_{CT} \times \eta_{Coll} = \eta_{Abs} \times IQE, \qquad (1.14)$$

while IQE is the number of electrons that reach the external circuit per absorbed photon. EQE_{PV} can be measured experimentally. In Fig. 1.11a the EQE_{PV} spectrum of a typical solar cell with active layer comprising PM6:Y6 (see Appendix for the details of the material systems and device fabrication) is plotted on left axis *versus* photon wavelength λ. The photocurrent density J_{Ph} generated by a device upon illumination over an extended spectral range under short circuit condition is determined by

$$J_{Ph} = q \int_0^\infty EQE_{PV}(\lambda) \times \phi(\lambda) d(\lambda) \qquad (1.15)$$

where $\phi(\lambda)$ is the spectral flux density of the light source. In the case of a standard solar cell measurements $\phi(\lambda) = \phi_{Sun}(\lambda)$ which is the air mass 1.5 (AM1.5G) solar spectrum with 100 mW/cm^2 power density. In Fig. 1.11a right axis, J_{Ph} has been

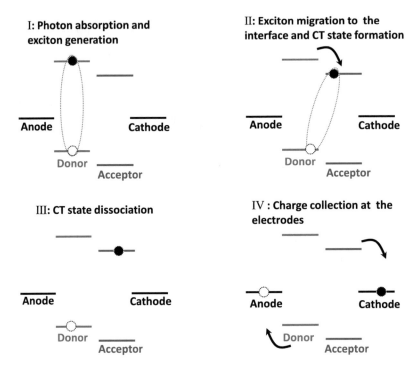

Fig. 1.10 A schematic picture of operational mechanism of BHJ organic solar cells

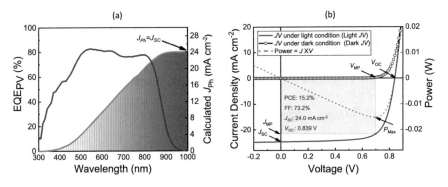

Fig. 1.11 **a** Right axis: EQE$_{PV}$ spectrum *versus* wavelength **a** Left axis: J_{SC} is calculated for the presented EQE$_{PV}$ spectrum. J_{SC} is the calculated value of the integral at the upper limit of the EQE$_{PV}$ spectrum **b** J-V curve of the PM6:Y6 device is plotted and the device characteristics are shown

calculated for the presented EQE$_{PV}$ spectrum. In a solar cell with moderate R_S at short circuit condition $J_{Ph} = J(V = 0)$, called the short circuit current (J_{SC}).

1.5.3 Power Conversion Efficiency

In order to perform as a power source, a solar cell must deliver electric power which is the product of a current and a voltage. The power conversion efficiency (PCE) of a solar cell is defined as the ratio of the solar cell maximum electrical power (maximum output power) and power of the incident light (input power) and it relates to the solar cell characteristics, as follows:

$$\text{PCE} = \frac{P_{Max\ Solar\ cell}}{P_{Sun}} = \frac{J_{SC}\, V_{OC}\, \text{FF}}{P_{Sun}}, \tag{1.16}$$

here V_{OC} is the open circuit voltage and FF is the fill factor. In Fig. 1.11b a typical current density *versus* voltage (J-V) curve of a solar cell is shown. The point where the electrical power reaches the maximum value is denoted as P_{Max}. The ratio of the maximum power output to the product of open circuit voltage and short circuit current is called the fill factor:

$$\text{FF} = \frac{P_{Max\ Solar\ cell}}{J_{SC}\, V_{OC}} = \frac{J_{Max}\, V_{Max}}{J_{SC}\, V_{OC}}. \tag{1.17}$$

1.6 Summary

In this chapter basic concepts of semiconductor physics were reviewed, the main differences in properties of organic and inorganic semiconductors were discussed and a general view of working principles of organic solar cells was presented. The next chapter will focus on different processes involving photocurrent and photovoltage generation in OSCs.

References

1. Kittel, C., & McEuen, P. (1976). *Introduction to solid state physics*. New York: Wiley.
2. Langmuir, I. (1919). The arrangement of electrons in atoms and molecules. *Journal of the American Chemical Society, 41*, 868–934.
3. Pauli, W. (1925). Über den Zusammenhang des Abschlusses der Elektronengruppen im Atom mit der Komplexstruktur der Spektren. *Zeitschrift Für Physik, 31*, 765–783.
4. Grundmann, M. (2010). *The physics of semiconductors: An introduction including nanoparticles and applications*. Berlin: Springer.
5. Würfel, P., & Würfel, U. (2016). *Physics of solar cells: From basic principles to advanced concepts*. New York: Wiley.
6. Meyers, H., & Myers, H. (1997). *Introductory solid state physics*. Boca Raton: CRC Press.
7. Ashcroft, N., Mermin, N. & et al. (1976). Solid state physics [by] Neil W. Ashcroft [and] N. David Mermin. New York: Holt, Rinehart.
8. Nelson, J. (2003). *The physics of solar cells*. Singapore: World Scientific Publishing Company.
9. Lambert, J. & Schneider, H. (1970). Physical organic chemistry. Wiley Online Library.

10. Demtröder, W. (2008). *Molecular physics: Theoretical principles and experimental methods.* New York: Wiley.
11. Bredas, J. (2014). Mind the gap! *Materials Horizons, 1,* 17–19.
12. Dennler, G., Lungenschmied, C., Neugebauer, H., Sariciftci, N., Latrèche, M., Czeremuszkin, G., & Wertheimer, M. (2006). A new encapsulation solution for flexible organic solar cells. *Thin Solid Films, 511–512,* 349–353.
13. Liu, Z., Li, J., & Yan, F. (2013). Package-free flexible organic solar cells with graphene top electrodes. *Advanced Materials, 25,* 4296–4301.
14. Yan, T., Song, W., Huang, J., Peng, R., Huang, L. & Ge, Z. 16.67.
15. Hashemi, D., Ma, X., Ansari, R., Kim, J., & Kieffer, J. (2019). Design principles for the energy level tuning in donor/acceptor conjugated polymers. *Physical Chemistry Chemical Physics, 21,* 789–799.
16. Takahashi, Y., Obara, R., Nakagawa, K., Nakano, M., Tokita, J., & Inabe, T. (2007). Tunable charge transport in soluble organic-inorganic hybrid semiconductors. *Chemistry of Materials, 19,* 6312–6316.
17. Eakins, G., Alford, J., Tiegs, B., Breyfogle, B., & Stearman, C. (2011). Tuning HOMO-LUMO levels: Trends leading to the design of 9-fluorenone scaffolds with predictable electronic and optoelectronic properties. *Journal of Physical Organic Chemistry, 24,* 1119–1128.
18. Haruk, A., Leng, C., Fernando, P., Smilgies, D., Loo, Y., & Mativetsky, J. (2020). Tuning organic semiconductor alignment and aggregation via nanoconfinement. *The Journal of Physical Chemistry C, 124,* 22799–22807.
19. Gledhill, S., Scott, B., & Gregg, B. (2005). Organic and nano-structured composite photovoltaics: An overview. *Journal of Materials Research, 20,* 3167–3179.
20. Clarke, T., & Durrant, J. (2010). Charge photogeneration in organic solar cells. *Chemical Reviews, 110,* 6736–6767.
21. Armin, A., Stoltzfus, D., Donaghey, J., Clulow, A., Nagiri, R., Burn, P., Gentle, I., & Meredith, P. (2017). Engineering dielectric constants in organic semiconductors. *Journal of Materials Chemistry C, 5,* 3736–3747.
22. Bässler, H. (1993). Charge transport in disordered organic photoconductors a Monte Carlo simulation study. *Physica Status Solidi (b), 175,* 15–56.
23. Marks, R., Halls, J., Bradley, D., Friend, R., & Holmes, A. (1994). The photovoltaic response in poly (p-phenylene vinylene) thin-film devices. *Journal Of Physics: Condensed Matter, 6,* 1379.
24. Knupfer, M. (2003). Exciton binding energies in organic semiconductors. *Applied Physics A, 77,* 623–626.
25. Sariciftci, N. & Heeger, A. (1994) Conjugated polymer-acceptor heterojunctions; diodes, photodiodes, and photovoltaic cells. Google Patents, US Patent 5,331,183.
26. Benson-Smith, J., Goris, L., Vandewal, K., Haenen, K., Manca, J., Vanderzande, D., Bradley, D., & Nelson, J. (2007). Formation of a ground-state charge-transfer complex in polyfluorene//[6, 6]-phenyl-C61 butyric acid methyl ester (PCBM) blend films and its role in the function of polymer/PCBM solar cells. *Advanced Functional Materials, 17,* 451–457.
27. Liu, Q., Jiang, Y., Jin, K., Qin, J., Xu, J., Li, W., Xiong, J., Liu, J., Xiao, Z., Sun, K. & et al. 18% Efficiency organic solar cells. *Science Bulletin,* 272–275.
28. Green, M., Dunlop, E., Hohl-Ebinger, J., Yoshita, M., Kopidakis, N., & Hao, X. (2020). Solar cell efficiency tables (version 56). *Progress In Photovoltaics: Research And Applications, 28.*

Chapter 2
Electro-optical Phenomena in Organic Solar Cells

Abstract As outlined in Chap. 1, charge photogeneration in organic solar cells is typically explained as a multi-step process. In this chapter, the details of each step, from light absorption to charge collection at the electrodes will be discussed from kinetics and energetics perspectives. Some theoretical models that have been used to describe the main phenomena will be explained. Furthermore, the competing processes at each step that can limit the overall efficiency of a device will be explored.

2.1 Light Absorption in Molecular Solids

2.1.1 Classical Point of View

From a classical point of view when an oscillating electric field $E(t) = E_0 \exp(-i\omega t)$ of electromagnetic radiation (light) hits a molecule (chromophore), it interacts with the molecular dipole moment which is an uneven distribution of positive and negative charges on the atoms. The effective centre of these distributions are in relative distance of x from one another so that the dipole moment is $\mu_d = qx$. An electric dipole oscillating in response to the electric field can be treated as a damped harmonic oscillator. As a result, $x = \frac{qE_0}{m_e} \frac{1}{(\omega_0^2 - \omega^2) - i\omega\gamma_d}$ where m_e is the mass of electron, ω_0 is the resonance frequency of the electric dipole, and γ_d is the damping constant. In a dielectric sample with N dipoles per unit volume the polarization density is given by $P = Nqx$. The electric displacement field in a dielectric is

$$D = \varepsilon_0 E_0 + P = \varepsilon_0 \left(1 + \frac{Nq^2}{\varepsilon_0 m_e} \frac{1}{(\omega_0^2 - \omega^2) - i\omega\gamma_d} \right) E_0, \tag{2.1}$$

where ε_0 is the vacuum permittivity and the term between the parenthesis is called the dielectric constant ε_r and it is the macroscopic measure for polarizability of a dielectric [1].

© The Author(s), under exclusive license to Springer Nature Switzerland AG 2022
N. Zarrabi, *Optoelectronic Properties of Organic Semiconductors*,
SpringerBriefs in Materials,
https://doi.org/10.1007/978-3-030-93162-9_2

The relation between the refractive index and the dielectric constant of a material can be derived from Maxwell's equation as: $\tilde{n} = \sqrt{\varepsilon_r}$ [2]. This raises two important points. Firstly similar to ε_r, the refractive index is frequency dependent and secondly, similar to ε_r, the refractive index of an absorbing media at each frequency is a complex number $\tilde{n} = n + ik$. Here n and k are the real and imaginary part of the refractive index, usually known as refractive index and extinction (attenuation) coefficient, respectively [3].

The propagation of an electromagnetic field of frequency ω in a medium with complex refractive index $\tilde{n}$ can be describe by a plane wave which has the electric field of the form $E(z, t) = E_0 \exp i (\bar{k} z - \omega t)$ where the wave vector $\bar{k} = \frac{2\pi \tilde{n}}{\lambda}$ and λ is the wavelength. This shows that the amplitude of the electric field ($E_0 \exp(\frac{-2\pi k}{\lambda} z)$) in an absorbing material decreases as the wave propagates along the z direction and so does the light intensity which has the form of

$$ I = \left| E_0 \exp \left(\frac{-2\pi k}{\lambda} z \right) \right|^2 = |E_0|^2 \exp \left(\frac{-4\pi k}{\lambda} z \right), \tag{2.2} $$

where $I_0 = |E_0|^2$ is the initial light intensity [4]. The light intensity in a medium empirically described by the Beer-Lambert law:

$$ I = I_0 \exp (-\alpha d). \tag{2.3} $$

Here α is the absorption coefficient of the media and d is the distance from the surface. By comparing to Eq. (2.2) it follows that

$$ \alpha = \frac{4\pi k}{\lambda}. \tag{2.4} $$

In cases where only the spectral shape of the absorption of a media is important a dimensionless parameter $A(\lambda)$ known as absorbance or optical density is commonly used instead of absorption coefficient and it is defined by

$$ I = I_0 10^{-A(\lambda)}. \tag{2.5} $$

Absorbance and absorption coefficient are related by $\alpha d = \ln 10 \cdot A$. In Fig. 2.1a, b and c, the n and k values of three well-known organic semiconductors, measured by spectroscopic ellipsometry (See Box 2.1) are shown. In Fig. 2.1d the absorption coefficient of the same material systems are plotted *versus* wavelength.

Box 2.1
Spectroscopic Ellipsometry: Ellipsometry is an optical technique for determining the optical properties (such as the refractive indices) of a dielectric thin film. In this technique a linearly polarized white light beam is directed onto

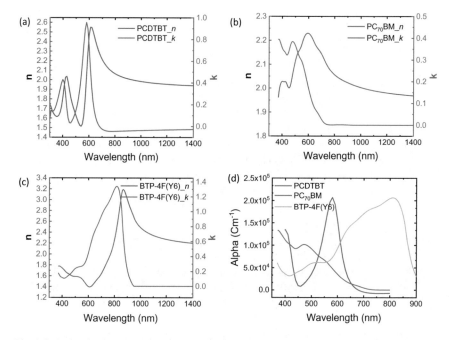

Fig. 2.1 Refractive index **n** and extinction coefficient **k** and absorption coefficient (α) of PCDTBT and PC$_{70}$BM and BTP-4F (Y6) are plotted *versus* wavelength

a sample at an incident angle of ϕ and the subsequent reflected beam from the sample is measured by passing through a rotating analyser followed by a detector.

Upon interaction with the sample, the reflected beam becomes elliptically polarized due to a phase change. The ratio between the Fresnel coefficient of the p-polarised (r_p) and s-polarised (r_s) component of the reflected light can be written as:

$$\rho = \frac{r_p}{r_s} = tan(\Psi) \exp(i\,\Delta).$$

Here Ψ represents the amplitude ratio and Δ represents the phase difference between P- and S-polarised light. A small changes in the thickness or optical constants of the film will change the phase difference.

Using ellipsometry, (Ψ,Δ) can be measured as a function of wavelength (200–1700 nm) for different angles of incidence (usually three angles are used in the range of 45–75°). (Ψ,Δ) are then used in an optical model to obtain refractive indices and thickness of the sample [5]. In the work described in this book, a J. A. Woollam M-2000 ellipsometer was used for the measurement and CompleteEASE 5.23 (J. A. Woollam) software was used for optical modelling.

Box 2.2

IQE Calculation: As mentioned in Sect. 1.5.2, charge generation efficiency in an OPV device is defined as the ratio of the number of electrons reaching the external circuit per incident photon which is referred to as the external quantum efficiency (EQE_{PV}). As a result, EQE_{PV} is a measure of both absorption and electronic properties of a device. On the other hand, internal quantum efficiency (IQE) is defined as the number charges extracted from a cell (reaches the external circuit) to the number of photons absorbed in the active layer which is the only layer in an OPV device responsible for charge carrier generation:

$$IQE = EQE_{PV}/\eta_{Abs}$$

In other words, IQE is solely a measure of the electronic properties of a device and can provide valuable information about the spectral dependency of charge generation and recombination in a cell.

In order to calculate the IQE an accurate measurement of the active layer absorption is needed. The active layer of an OPV device is typically sandwiched between a stack of different materials being interlayers and two electrodes. As a result, the optical field distribution and the absorption in the active layer are highly affected by the cavity interference effects induced by the reflective electrodes and also the parasitic absorption of non-active layers.

Transfer matrix modelling is a powerful method that can simulate the electric field distribution and parasitic absorption (*PA*) in a device stack [6]. In this method, each layer of the device is defined based on its complex refractive index $\tilde{n} = n + ik$ (measured by ellipsometry) and its thickness. From this analysis, it can describe the transmitted and reflected component of the electric field at each layer from which the optical filed distribution and the absorption of each layer in the whole stack can be evaluated [7]. In order to account for the effect of optical scattering from the device (which can not be taken into account in the transfer matrix simulation) the total device reflectance (R) must be measured experimentally. In the work described in this book, the near-normal incidence reflectance of the devices were recorded using either a PV Measurements Inc.QEX7 setup or a universal reflectance attachment (URA) on a Perkin-Elmer Lambda 950 spectrophotometer. Finally, the IQE can be calculated as:

$$IQE = \frac{EQE_{PV}}{1 - R - PA}$$

2.1.2 Quantum Mechanical Point of View

From the perspective of quantum theory, the state of a molecular system is a combination of the state of the electrons and the nuclei. The electronic state is described by the electronic wave function ψ_{el} which depends on the position of electrons and the nucleus. Schematically, the electronic states are shown by horizontal lines (see Fig. 2.2a) where S_0 is the the ground state, S_1 is the first excited state and so on. If nuclei are considered frozen, and electron-electron interaction neglected (otherwise an analytical solution cannot be achieved), the energy of an electronic state can be approximated by the molecular orbitals. This is only a rough estimation, however the energy difference between the S_0 and S_1 is conventionally shown as the difference between HOMO and LUMO. In molecular solids, these energies are respectively analogous to ionization potential (*IP*) and electron affinity (*EA*) that can be experimentally measured.

In reality, the oscillation of the nucleus in a molecular system should also be considered to find the energy state of the whole molecular system. The vibrational state is described by ψ_{vib}. In this picture the energy of a molecule is calculated based on the different position of each nucleus R_i. This is analogous to a system of pendulums coupled by springs. To describe such a system, one should write a set of differential equations which describe the motion of each nucleus while it is coupled to all the other nuclei in the system. This can mathematically best be treated by defining a set of normal mode coordinates q_i (nucleus displacement coordinates) from the nuclear coordinates R_i (Cartesian coordinates). A potential surface describes the energy of the system for each q_i. The potential that each nucleus experiences can be approximated by a harmonic oscillator with vibrational frequency of ω and associated vibrational energy of $E_n = (n + \frac{1}{2})\hbar\omega$ (see Fig. 2.2b). It is notable that, since the nuclei are much heavier than the electrons, the nuclear wavefunctions are often considered to be independent of the electronic wavefunctions and hence electron-nuclei interactions are neglected. This so-called Born–Oppenheimer approximation is necessary for the quantum chemical calculation to be computationally viable [8].

A photon can be absorbed by a molecule if the energy of the photon is greater or at least equal to the energy gap of that molecule. By absorbing a photon, the total energy of the molecule increases and the molecule transitions from the initial state Ψ_i, so called the ground state, to the final state Ψ_f, so called the exited state. The rate of the transition k_{if}, is described by the Fermi's golden rule as

$$k_{if} = \frac{2\pi}{\hbar}|\langle\Psi_i|H'|\Psi_f\rangle|^2\rho, \tag{2.6}$$

where ρ is the density of final states and H' is the perturbation Hamiltonian. Using the Born–Oppenheimer approximation, the total state of a molecule can be written as the product of the electronic and vibrational state where the Hamiltonian which is a dipole operator $\bar{\mu} = q\hat{r}$ can only act on the electronic state

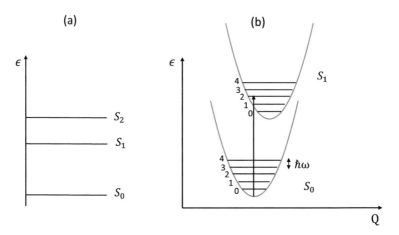

Fig. 2.2 **a** Energy level diagram of electronic states showing the ground state and the first two excited state. **b** Energy level diagram including vibrational energy states and the transition from the zeroth vibrational level of the ground state to the second vibrational level of the first excited state

$$k_{if} = \frac{2\pi}{\hbar} |\langle \psi_{el,f} | \bar{\mu} | \psi_{el,i} \rangle|^2 |\langle \psi_{vib,f} | \psi_{vib,i} \rangle|^2 \rho. \tag{2.7}$$

The transition rate corresponds to the number of photons absorbed per second per molecule as the result, the magnitude of the absorption coefficient is proportional to the transition rate. The first term here known as the electronic coupling and is dependent on the overlapping of the initial and final states. The larger the electronic coupling the higher is the transition rate. The overall intensity of the absorption spectrum is dictated by the electronic coupling. The second term is known as the Franck–Condon constant and it describes the probability of the transition from zeroth vibrational level of the ground state to the nth vibrational level of the excited state. The shape of the absorption spectra is known to be governed by the Frank-Condon constant.

2.2 Exciton

2.2.1 Generation and Migration

Absorption of a photon via the above discussed dipolar transition by a molecule results in the formation of an excited state on that molecule. As mentioned before the excited state which is an electrically neutral quasi particle is conventionally referred to as an exciton.

The excitation energy can migrate from one molecule to another molecule via an incoherent process known as energy transfer or Förster transfer. Energy transfer occurs when the emission spectrum of one molecule called the excitation donor overlaps with the absorption spectrum of another molecule called the excitation acceptor. Energy transfer is a non-radiative process i.e. there is no photon emission or absorption involved and the donor and acceptor are coupled via dipole-dipole interactions. The rate of energy transfer is given by

$$k_{ET} = \frac{1}{\tau_D} \left(\frac{R_0}{r} \right)^6, \tag{2.8}$$

where R_0 is the Förster distance and describes the spectral overlaps, r is the distance between the donor and the acceptor and τ_D is the exciton lifetime. The average distance an exciton may travel during its lifetime is defined as exciton diffusion length L_{diff} which is defined as

$$L_{diff} = \sqrt{2mD\tau_D}, \tag{2.9}$$

where D is the diffusion constant and m is the dimensionality of the space wherein the exciton diffuses. Typical diffusion lengths in organic semiconductors are less than 20 nm.

As mentioned before, excitons can be described as an electron and hole pair that are bound with Coulombic attraction. The exciton binding energy in organic semiconductors is on the order of 0.3–1 eV [9, 10]. To have free electron and hole which is essential to produce photocurrent in a solar cell the exciton must be dissociated.

Exciton dissociation probability is very low in a neat organic semiconductor and most excitons recombine, either radiatively or non-radiatively, within 100 ps–1 ns. This recombination is mostly radiative in organic semiconductors which are optimised for light emission. Such materials are currently used in state-of-the-art displays and televisions. Whilst such organic light emitting diodes (OLED) operate very efficiently, the old statement that "a good solar cell should be a good LED and vice versa" [11] does not apply here due to the excitonic nature of pristine organic semiconductors. As a result, the very first solar cells, made from a single layer of polymer as the active layer, failed due to very low charge generation efficiency. In 1986, Tang reported that the charge generation efficiency can be improved in a bilayer system (an organic heterojunction), if there is an *EA* difference between the layers material systems [12]. Exciton dissociation in this scenario occurs at the interface of the photon absorber i.e. the donor, and an electron acceptor. The difference in the *EA* of the donor and the acceptor at the interface creates a driving force that can dissociate the exciton. In order to have efficient exciton dissociation the thickness of the layer of the donor material system must be on the order of exciton diffusion length so that the excitons can reach the interface prior to recombination. In 1995, the ultimate active layer structure of bulk heterojunction (BHJ) in which the donor and acceptor are mixed at the nanometer scale was introduced. This structure of the active layer ensures that domain sizes comparable to exciton diffusion lengths in organic semiconductors are feasible [13].

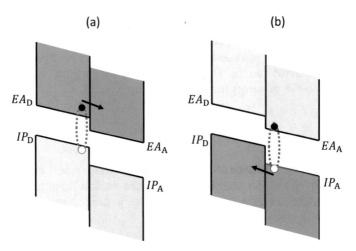

Fig. 2.3 The two pathways of charge generation **a** Channel 1, the photoexcitation happens in the donor and electron transfers from the donor LUMO to acceptor LUMO and **b** Channel 2, the photoexcitation happens in the acceptor and hole transfers from the acceptor HOMO to the donor HOMO

Knowledge of exciton diffusion lengths for a specific material system can be beneficial to optimise the domain sizes and therefore maximise the number of dissociated excitons. The measurement of diffusion lengths in organic semiconductors has been a consistent challenge [14–17]. In Chap. 3, we will discuss more about this matter and we present our own developed method of exciton diffusion length measurement.

The charge generation process in an organic solar cell is often explained and understood based on a scenario where the photoexcitation and exciton generation happens in the electron donor material. This is because, historically, these devices were mostly made from a polymer donor and a fullerene acceptor where there was a substantial overlap between the absorption of the two with the polymer having relatively much higher absorption coefficient. As a result, the donor photoexcitation governed the charge generation. Photoexcitation can also happen in the electron acceptor. This is particularly important in the active layers where the acceptor is the majority component or the acceptor has a spectrally distinct absorption. In this scenario the driving force for exciton dissociation is the *IP* difference between the two material systems which promotes hole transfer from the excited electron acceptor (hole donor in this case) to the electron donor (hole acceptor). The two charge generation pathways are known as Channel 1 and Channel 2 and they are summarized in Fig. 2.3 [18, 19].

2.2.2 Dissociation and Charge Transfer State Formation

Once an exciton reaches the interface of the donor and the acceptor, it can be subjected to charge transfer in which a slightly less strongly bound electron and hole pair called the charge transfer (CT) state can be formed. The theoretical description of the rate of charge transfer mechanism can be explained by Marcus theory.

Marcus theory was first established to describe the charge transfer reaction between ions in a solution, where the electron transfer between the reactant and the product happens without chemical bond breaking in the molecules and it only cause a reorganization in electrostatic configuration of the ions and the environment (the solvent) [20]. Later, it has been shown that it can also be used to describe photoinduced charge transfer in organic semiconductor blends [21, 22].

When electron transfers from one site to another, electron moves in the actual potential landscape of the nucleus. The Franck–Condon principle states that the movement of the electron is so fast compared to the nuclei that the nuclei effectively does not move during the charge transfer process. However, the molecules must reorganised their configuration in such a way that the extra charge can be stabilised in its new coordinates. The amount of energy that is needed for the reconfiguration is called the charge transfer reorganisation energy λ_{DA}. In the picture of Marcus theory, the reactant (excited donor molecule and the acceptor (in the ground state) denoted as D^*A) and the product (positively charged donor and the negatively charged acceptor D^+A^-) energy states are shown with two potential energy surfaces as shown in Fig. 2.4a with the equilibrium configuration of q_1 and q_2. Charge transfer can only happen when the energy of the reactant and the product are identical (energy conservation must satisfy). This happens at the intersection of the two parabolas and the energy that the system must overcome (activation barrier) to reach this point is G_B. The activation energy can be mathematically evaluated from the energy difference between surfaces minima ΔG^0 and λ_{DA} as $G_B = \frac{(\lambda_{DA}+\Delta G^0)^2}{4\lambda_{DA}}$.

In thermal equilibrium, the transfer rate $k_{A\rightarrow D}$ can be determined using Boltzmann statistics:

$$k_{A\rightarrow D} = A \exp\left(\frac{-G_B}{k_B T}\right) = A \exp\left(-\frac{(\lambda_{DA} + \Delta G^0)^2}{4\lambda_{DA} k_B T}\right). \tag{2.10}$$

It can be inferred that in Marcus picture the difference between $-\Delta G^0$ and λ_{DA} is a key parameter.

For a constant λ_{DA}, as $-\Delta G^0$ increases $k_{A\rightarrow D}$ also increases until the maximum rate reached at $\lambda_{DA} = -\Delta G^0$. (see Fig. 2.4b) At this point the reaction has no activation barrier. Further increase of the $-\Delta G^0$ again will decrease the transfer rate. This is known as Marcus inverted region (see Fig. 2.4c) [23].

ΔG^0 is known as the driving force of the transfer process and it is the difference in Gibbs free energy between the initial and final states, $\Delta G^0 = \Delta H - T\Delta S$, with H being the enthalpy and ΔS is the change in the entropy and T, as always, is the temperature.

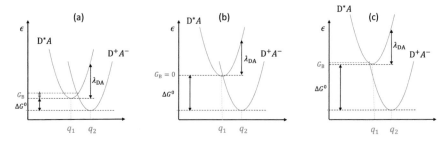

Fig. 2.4 a Energy parabolas of excited donor:acceptor blends and positively charge donor:negatively charged acceptor. **b** Energy parabola showing the maximum transfer rate $G_B = 0$. **c** Marcus inverted region

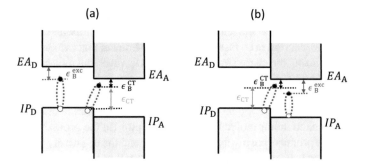

Fig. 2.5 Schematic energy diagram for the driving force ΔG^0 evaluation **a** for donor photoexcitation and **b** for acceptor photoexcitation

In solid state BHJ organic solar cells, ΔG^0 is the difference between the initial exciton, of the donor or the acceptor, and the CT states energy at the interface. As it can be seen in Fig. 2.5, in the scenario where the photoexcitation happens in the donor, $\Delta G^0 = (EA_A - IP_D - \epsilon_B^{CT}) - (EA_D - IP_D - \epsilon_B^{exc}) = (EA_A - EA_D) - (\epsilon_B^{CT} - \epsilon_B^{exc})$ and similarly in the scenario where the photoexcitation happens in the donor $\Delta G^0 = (IP_A - IP_D) - (\epsilon_B^{CT} - \epsilon_B^{exc})$.

Marcus theory has been widely used to describe charge transfer in organic semiconductors. The impact of the driving force has been investigated in many material systems. For example, Ward et al. [24] have conducted time resolved photoluminescence measurements on series of donor-acceptor blends in which (using the same donor) the EA of the acceptor were varied sequentially. Their results show a correlation between charge transfer rate and the EA's offset. Most importantly, they reported that in their examined material systems, for $\Delta G^0 = -0.4$ eV the charge transfer rate is maximal and it decreases significantly above and below the optimal value which is in agreement with Marcus theory (forward and inverse Marcus regime).

2.3 Charge Transfer State

2.3.1 Dissociation

Despite locating on different adjacent molecules, the electron and hole in CT states are in a moderate distance from one another which means that they are bound together *via* Coulomb attraction. Thus similar to excitons, electron and hole in CT states are subjected to recombination. The CT state recombination is known as geminate recombination since the charges are initially formed from the same exciton. This process can happen either radiatively or non-raditively.

Dissociation probability of CT states was first described by Onsager theory which was originally developed to describe the probability of photogenerated electron-hole pair dissociation in a weak electrolyte [25]. Onsager theory states that right after excitation an energetically hot electron (with excess thermal energy) will be created. The hot electron thermalises after a short while at the distance a, known as the thermalisation length, from the hole. The competition between dissociation and recombination at this point is dependent on the amount of Coulomb attraction affecting the electron. For a medium with effective dielectric constant ε_r an electron can escape from a hole potential if the Coulomb attraction is less than the environment thermal energy $k_B T$. The distance at which the two energies are equal is known as Coulomb capture radius r_C

$$ r_C = \frac{q^2}{4\pi \varepsilon_r \varepsilon_0 k_B T}. \tag{2.11} $$

Onsager theory predicts that if a is greater that the r_C then electron and hole can be considered as fully dissociated. However, If a is smaller than r_C the probability of dissociate is defined as $P(E)$ while the probability of geminate recombination would be $1 - P(E)$ where E is the strength of any applied electric field. It has been observed that describing the field dependent photo-generation yield in organic solids using Onsager theory is not plausible since it requires a thermalisation length of $\sim$16 nm which is much larger than nearest-neighbor distance (less than 1 nm).

In 1984 Braun presented a modified model based on Onsager theory in which he considered a finite lifetime for CT states in solids. The CT states are depleted either by geminate recombination with corresponding rate constant of k_f (either radiatively or non-radiatevly) or by dissociation to separated charges (CS) with corresponding (electric filed dependent) rate constant of $k_d(E)$ [26]. The dissociation probability (dissociation quantum yield) in Braun's model is defined based on the competition between geminate recombination rate and the dissociation rate:

$$ P(E) = \frac{k_d(E)}{k_d(E) + k_f}. \tag{2.12} $$

The most important consideration in Braun's model is that CT state dissociation to CS is a reversible process meaning that CT states can be recreated by separated charges. The electric field dependent dissociation constant is defined as:

$$k_d(E) = \frac{e(\mu_p + \mu_n)}{\varepsilon_r \varepsilon_0} \left(\frac{3}{4\pi \mathbf{a}^3} \exp\left(\frac{-\Delta\epsilon_b}{k_B T} \right) \right) \left[1 + b + \frac{b^2}{3} + \frac{b^3}{18} + \cdots \cdots \right].$$
(2.13)

The first term is the Langevin recombination rate constant where μ_n and μ_p are the electron and hole mobilities. The second term is a Boltzmann distribution that describes the static dissociation rate without the contribution of the electric field ($\mathbf{a}$ is the thermalisation length and $\Delta\epsilon_b$ is the binding energy following thermalisation). The final term is the approximation of a first-order Bessel function where $b = q^3 E / 8\pi \varepsilon_r \varepsilon_0 k_B^2 T^2$ and it describes the increment of the dissociation rate due to presence of electric field (E).

2.4 Charge Collection

Following CT state dissociation the photogenerated free electron and hole must be collected at their respective electrodes, being the cathode and the anode, to contribute to photo-current. The efficiency of charge collection η_{CC} is defined as the ratio of the extracted charges at the electrode to the photogenerated charges. η_{CC} is essentially determined by the competition between charge extraction and recombination of free carriers. Free carriers are subjected to bimolecular recombination, trap-assisted recombination within the bulk and surface recombination at the electrodes.

2.4.1 Charge Transport

As discussed previously, due to the disordered nature of organic semiconductors, charge transport in these material systems is governed by hopping transport between the adjacent localised sites (molecules or different segment of a polymer). This conduction mechanism is a combination of thermally activated hops and tunneling between localized states. Similar to electron transfer at the donor-acceptor interface, the transfer rate of the free electrons (holes) in the acceptor (donor) phase from site i to site j can be described by nonadiabatic Marcus transfer rate [20, 27]:

$$k_e^{ij} = \frac{2\pi}{\hbar} \frac{|V_{ij}|^2}{\sqrt{4\pi \lambda_{CS} k_B T}} \exp\left(-\frac{(\Delta\epsilon_{LUMO}^{ij} + \Delta U^{ij} + \lambda_{CS})^2}{4\lambda_{CS} k_B T} \right),$$
(2.14)

where V_{ij} is the electronic coupling between the two sites, $\Delta\epsilon_{LUMO}$ is the difference in the LUMO level energy of the two sites, ΔU^{ij} is the difference in the Coulomb potential between the two charge configurations and, λ_{CS} is the reorganization energy due to the separated charge hopping.

The drift mobility of charge carrier, μ, is evaluated from the Einstein relation

$$\mu = \frac{eD}{k_B T}.$$ (2.15)

D is the charge diffusion coefficient which can be evaluated from

$$D = \frac{1}{2m} \sum_l r_l^2 k_l P_l.$$ (2.16)

Here m is the dimensionality of the space and l is a specific transfer pathway, r_l is distance between two site within l and k_l is the corresponding transfer rate that can be calculated from Eq. (2.14), and P_l is the relative probability of charge transfer within l which can be calculated from $P_l = k_l / \sum_l k_l$ [28, 29].

2.4.2 Bimolecular Recombination

The direct recombination between free electrons and holes occurs *via* bimolecular recombination. This process follows a second-order recombination kinetics which depends on the concentration of both electrons and holes. The recombination rate is given by

$$R = \beta_{Bulk} np$$ (2.17)

where β_{Bulk} is the bulk recombination rate constant. Bimolecular recombination in a material with low mobility can be generally described by the Langevin recombination rate which assumes that the effective recombination rate is determined by the encounter rate of free carries when the Coulomb radius is larger than the hopping distance. Since this process is diffusion-limited, it is dependent on the mobility of the free charge carriers. It has been shown that the recombination rate in pristine organic semiconductors can be more or less explained by the Langevin recombination rate [30] with a rate constant of

$$\beta_L = \frac{q(\mu_p + \mu_n)}{\varepsilon_r \varepsilon_0},$$ (2.18)

where μ_n and μ_p are the electron and hole mobility. Conversely, the experimentally determined bimolecular recombination rate constants in donor-acceptor bulk heterojunctions, β_{Bulk}, is generally smaller than the value expected from the Langevin relation. In the donor-acceptor BHJ the recombination of free electrons and holes can only happen at the interface thus the suppressed recombination rate was initially attributed to the geometrical separation (confinement) of electrons and holes within their respective domains. On the other hand, since electrons and holes do not have same mobilities and are confined in separated domains, it has also been suggested that different arrival times of the electron and hole to the donor-acceptor interface may be responsible for the suppressed recombination rate. This "geometrical" suppression of recombination can be formulated as

$$\gamma_{Geo} = \frac{\beta_{en}}{\beta_L},$$ (2.19)

where γ_{Geo} is the geometrical reduction factor and β_{en} is the encounter rate constant. In such cases, the more imbalanced the electron and hole mobility, the more reduced the encounter rate becomes as the faster carrier must wait at the interface to meet the slower carrier [31, 32]. Kinetic Monte Carlo simulations for realistic domain sizes has revealed that this geometrical effect only becomes relevant when the domain sizes are extremely large (>30 nm) which is not generally the case in BHJ OSCs [33]. For typical phase separations of 5–10 nm, in turn, it was found that the geometrical confinement of electrons and holes in their respective domains only suppresses the bimolecular recombination by a factor of 10 at most. Few exceptional material systems, however, exhibit strongly reduced recombination (by factor of 100–2000). This further reduction relative to Langevin recombination rate constant can be defined as

$$\gamma = \frac{\beta_{Bulk}}{\beta_L}$$ (2.20)

In general the generation and recombination of electrons and holes in BHJs occurs *via* CT states, acting as intermediate charge recombination/generation centres. In Fig. 2.6c a complete overview of processes involved in charge generation are shown in a schematic state diagram. Once an electron and a hole encounter each other at the interface (with the rate constant β_{en}) they form a CT state. The CT state either recombines to ground state with the rate constant k_f or dissociates again to free carriers with the rate constant k_d.

The probability of CT state dissociation is described by Eq. (2.12). The correlation between the charge generation quantum yield and the reduction factor has been confirmed, experimentally, by Shoaee et al [34]. The authors have shown that the reduction factor depends on the kinetics of the CT states and more specifically on the ratio between dissociation rate constant of the CT states to free carriers k_d as well as the decay rate constant of the CT state to the ground state k_f. However, in systems with strongly reduced recombination, it is not yet known whether the reduction is due to a suppressed k_f or improved k_d. We will return to this matter in Chap. 5.

2.4.3 Trap-Assisted Recombination

Free electrons and holes can also recombine through traps. Traps are energy states (ϵ_t) within the energy gap caused by impurities and imperfections in the semiconductor matrix or by the extension of the density of states into the gap. A trap state can be occupied by an electron or a hole. Electron traps also known as acceptor-like traps are

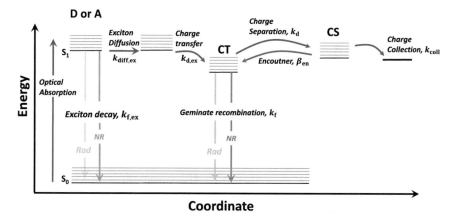

Fig. 2.6 State diagram describing the generation and recombination processes in a BHJ OPV device. Photoexcitation creates excitons in donor or acceptor (S1) which either dissociate at the donor:acceptor interface to form interfacial charge transfer (CT) states or recombine (radiatively or non-radiatively) to the ground state. The interfacial CT states either dissociate to free carriers or recombine (radiatively or non-radiatively) to the ground state. Free carriers are either collected at the electrodes and contribute to the photocurrent or recombine and reforms the CT state

negatively-charged when occupied by an electron and neutral when empty (occupied by a hole). Hole traps, also known as donor-like traps on the other hand, are neutral when occupied with an electron and positively-charged when occupied by a hole (i.e. when empty). The description of generation and recombination *via* trap states was first developed by Shockley, Read and Hall [35, 36]. In Fig. 2.7 a schematic picture of trap-assisted generation and recombination statistics *via* electron traps is shown. The net recombination rate for electron is given by $U_e^- = R_n^- - G_n^-$ and the net recombination rate for holes is given by $U_h^- = R_p^- - G_p^-$. Here, $R_n^- (R_p^-)$ is the capture rate of electrons (holes) into traps and $G_n^- (G_p^-)$ is the escape rate of electrons (holes) from the traps and can be defined as:

$$R_n^- = C_n n N_t (1 - f(\epsilon_t)) \quad \text{and} \quad G_n^- = C_n n_1 N_t f(\epsilon_t), \quad (2.21)$$

Fig. 2.7 Schematic of energy state diagram of trap-assisted generation and recombination statistics *via* electron traps. $R_n^- (R_p^-)$ is the capture rate of electrons (holes) into electron traps and $G_n^- (G_p^-)$ is the escape rate of electrons (holes) from the electron traps

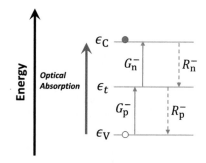

$$G_p^- = C_p p_1 N_t (1 - f(\epsilon_t)) \quad \text{and} \quad R_p^- = C_p p N_t f(\epsilon_t). \tag{2.22}$$

Furthermore, N_t is the density of trap states at energy level ϵ_t, $f(\epsilon_t)$ is the probability for an electron to occupy a trap state at energy ϵ_t, $n_1 = N_C \exp\left(\frac{\epsilon_t - \epsilon_C}{k_B T}\right)$ and $p_1 = N_V \exp\left(\frac{\epsilon_V - \epsilon_t}{k_B T}\right)$ are the concentration of electrons in the conduction band and holes in the valence band in the case the Fermi level coincides with the trap state energy and, C_n (C_p) is the electron (hole) capture coefficient. At steady state, the net recombination rate between electrons and holes becomes

$$U_{SRH}^- = R_n - G_n = R_p - G_p = \beta_{SRH} n p, \tag{2.23}$$

from which the $f(\epsilon_t)$) and $1 - f(\epsilon_t)$ can be calculated and consequently the trap-assisted recombination rate constant β_{SRH} can be written as:

$$\beta_{SRH}^- = \frac{C_n C_p N_t}{C_n(n + n_1) + C_p(p + p_1)} \tag{2.24}$$

An analogous expression can be obtained for generation and recombination *via* hole traps [37].

2.4.4 Surface Recombination

In order to contribute in the photocurrent, free charge carriers must be collected at the right electrode. In other words, the electrodes must be selective for the right type of carriers. Surface recombination is defined as the free charge carrier collection at the wrong electrode, that is, holes at the cathode and electrons at the anode [38]. In order to prevent charge carriers from recombining at the electrode interface, additional layers between the active layer material and the metal or transparent electrode are employed in the solar cell structure (See Fig. 1.8). For all presented data and calculations in this thesis, the effect of surface recombination is assumed to be negligible.

2.5 Theoretical Limit of the Power Conversion Efficiency of a Solar Cell

Similar to all power generating devices, a solar cell has a theoretical limit of power conversion efficiency. In 1961, Shockley and Queisser (SQ) developed a model to calculate the efficiency limit for a single layer p-n junction solar cell based on the thermodynamic principle of detailed balance [39]. The original SQ model is based on highly simplified assumptions, however the main idea of detailed balance in the model is the cornerstone of understanding the power conversion efficiency in the field. The limiting factors of the power conversion efficiency can be explained in the following steps:

Photon absorption and thermalisation: A photon can be absorbed by a solar cell if its energy is higher than the bandgap $\epsilon > \epsilon_g$ and the solar cell is transparent to the photons with energies lower than the bandgap. As a result, the first step of energy loss is due the spectral transparency of a solar cell. Also the excitation to high energy levels ($\epsilon > \epsilon_g$) will be thermalised within a short time scale to the energy of the electronic gap through which conduction happens. The second step of the energy loss would be *via* thermalisation of excited states. SQ assumes that the absorptivity is a step function which is 1 for $\epsilon > \epsilon_g$ and it is 0 for $\epsilon < \epsilon_g$. However, this can never be satisfied in real solar cells. The absorptivity is always less than 1 for $\epsilon > \epsilon_g$ due the finite thickness and finite absorption coefficient of solar cells. Also due to energetic disorder the solar cell always absorbs photons with energies bellow the bandgap. In organic BHJ OSCs, the photoexcitation of the donor and the acceptor thermalises into the CT states which technically makes it the effective energy gap of the OSC. The SQ model also assumes that for each photon with energy $\epsilon > \epsilon_g$ one electron-hole pair will be collected. However, absorption of photons can happen in the inter-layers which does not contribute to the photocurrent. As a result, the photocurrent of a solar cell is best described by

$$J_{Ph} = q \int_{\epsilon_{min}}^{\infty} EQE_{PV}(\epsilon) \times \phi_{sun}(\epsilon)d(\epsilon) \tag{2.25}$$

which is analogous to Eq. (1.15) (written in terms of photon energy). The lower integration limit ϵ_{min} is ideally zero but in practice given by the lower limit of the EQE_{PV} measurement. If optical loss was the only loss mechanism in the system then $qV_{OC} = \epsilon_g$.

Radiative recombination of charge carrier: Detailed balance states that in thermal equilibrium all microscopic processes are exactly equilibrized with their respective reverse process. Utilizing this, Shockley and Queisser inferred that in thermal equilibrium (the device has the same temperature as the environment and there is no external energy source) the rate of absorption and relative recombination must be the same. As a result, a recombination current also known as the dark saturation current due to radiative recombination of electron and holes of the solar cell is given by

$$J_0^{Rad} = q \int_{\epsilon_{min}}^{\infty} EQE_{PV}(\epsilon) \times \phi_{BB}(\epsilon)d(\epsilon) \tag{2.26}$$

where $\phi_{BB}(\epsilon)$ is the spectral flux density of the black body (environment) at room temperature and is given by

$$\phi_{BB}(\epsilon) = \frac{2\pi\epsilon^2}{h^3c^2} \frac{1}{[\exp(\frac{\epsilon}{k_BT}) - 1]} \tag{2.27}$$

where c is the speed of light.

If radiative recombination is the only process contributing to the dark current then the V_{OC} is

$$V_{OC}^{Rad} = \frac{kT}{q} \ln\left(\frac{J_{Ph}}{J_0^{Rad}} + 1\right) \qquad (2.28)$$

This is known as the radiative limit of the open circuit voltage. In this scenario $V_{OC} = V_{OC}^{Rad}$ and the $\Delta V_{OC}^{Rad} = \epsilon_g - V_{OC}^{Rad}$ would be the open circuit voltage loss due to radiative recombination.

Non-radiative recombination: The actual recombination current in a solar cell is the sum of radiative and non-radiative recombination currents $J_0 = J_0^{Rad} + J_0^{NR}$. As the result the actual V_{OC} of a solar cell is usually much lower than the V_{OC}^{Rad}. In [11], Rau showed that the electroluminescence quantum efficiency (EQE$_{LED}$) which is defined as as ratio of the number of photons emitted from the device (performing as an LED) to the number of electrons injected into the device, can be used as a measure to calculate non-radiative losses in open circuit voltage. In his paper, he showed that in the dark ($J_{sc} = 0$) by applying a voltage V the injected current would be equal to $J_{inj} = J^{Rad}(V) + J^{NR}(V)$ and thus

$$EQE_{LED} = \frac{J^{Rad}}{J_{inj}}, \qquad (2.29)$$

here $J^{Rad} = J_0^{Rad}[\exp(qV/k_BT) - 1] \approx J_0^{Rad}[\exp(qV/k_BT)]$. At applied voltage $V = V_{OC}$ the total current is equal to zero and thus $J_{inj}(V_{OC}) = J_{SC}$. As a result Eq. (2.29) can be written as

$$\ln(EQE_{LED}) = \ln\left[\frac{J_0^{Rad}[\exp(qV_{OC}/k_BT)]}{J_{inj}(V_{OC})}\right] = (qV_{OC}/k_BT) + \ln\left[\frac{J_0^{Rad}}{J_{SC}}\right]. \qquad (2.30)$$

Using Eq. (2.28) the non-radiative loss of the V_{OC} is given by:

$$\Delta V_{OC}^{NR} = V_{OC}^{Rad} - V_{OC} = -\frac{k_BT}{q}\ln(EQE_{LED}). \qquad (2.31)$$

where EQE$_{LED}$ can be measured experimentally.

2.5.1 Charge Transfer State as the Effective Energy Gap of an Organic Solar Cells

If charge generation and recombination occur *via* intermediate CT states then the effective energy gap of the system would be the CT state. This means that the energy of the CT state defines the upper limit of the V_{OC} [40, 41]. As a result, the accurate determination of the CT state energy would be necessary for the V_{OC} losses calculation.

The standard expression for non-adiabatic charge transfer between one donor D and one acceptor A is given by [42]:

$$k_{D \to A} = \frac{2\pi}{\hbar} \frac{|V_{DA}|^2}{\sqrt{4\pi k_B T \lambda_{DA}}} \exp\left(-\frac{(\lambda_{DA} + \Delta\epsilon_{DA})^2}{4\lambda_{DA} k_B T}\right). \tag{2.32}$$

where V_{DA} is the electronic coupling determined by the overlap of the electronic wavefunction of the donor and the acceptor and $\Delta\epsilon_{DA}$ is the difference between initial and final states. The electron transfer is considered to be non-adiabatic when the electronic coupling is weak.

By considering the CT state ground state and the excited state as the initial and the final states in the transition the transfer rate can be written as:

$$k_{CT} = \frac{2\pi}{\hbar} \frac{|V_{CT}|^2}{\sqrt{4\pi k_B T \lambda_{CT}}} \exp\left(-\frac{(\epsilon_{CT} + \lambda_{CT} - \epsilon)^2}{4\lambda_{CT} k_B T}\right), \tag{2.33}$$

where ϵ_{CT} is the energy difference between the initial and final state and λ_{CT} is the reorganisation energy and ϵ is the photon (excitation) energy. It has been shown that, considering the transfer rate Eq. (2.33), the absorption cross section of the CT state at energy ϵ can be described by

$$\sigma(\epsilon)\epsilon = \frac{f_\sigma^{CT}}{\sqrt{4\pi k_B T \lambda_{CT}}} \exp\left(-\frac{(\epsilon_{CT} + \lambda_{CT} - \epsilon)^2}{4\lambda_{CT} k_B T}\right). \tag{2.34}$$

The absorption coefficient α in the spectral region of the CT state absorption is equal to $N_{CT}\sigma$ where N is the number of CT state and σ is the absorption cross section. As a result,

$$\alpha(\epsilon) = \frac{f_\alpha^{CT}}{\epsilon\sqrt{4\pi k_B T \lambda_{CT}}} \exp\left(-\frac{(\epsilon_{CT} + \lambda_{CT} - \epsilon)^2}{4\lambda_{CT} k_B T}\right). \tag{2.35}$$

This expression can be used to extract CT state parameters from α. In Fig. 2.8 we used Eq. (2.35) to obtain CT state parameter in a PCDTBT:PC$_{70}$BM blend. Note that $f_\alpha = N_{CT} f_\sigma$. CT states usually have very low absorption cross sections thus, it is challenging to measure absorption coefficients for these states. In many studies the EQE$_{PV}$ spectrum has been used alternatively. However, the spectral shape of EQE$_{PV}$ is subjected to thickness dependent cavity interference effects within a device. A new procedure has been proposed by Kaiser et al. [43] to extract a better estimation for α in sub-gap energy regions. In this method, the EQE spectra of devices with different active layer thicknesses (same material system) are fitted simultaneously for one extinction coefficient using an iterative transfer matrix. In the work described in this book this method has been used whenever an accurate value of the CT state parameters was needed.

Fig. 2.8 Absorption coefficient in CT state spectral range and the CT state parameters obtain from the fit to Eq. (2.35)

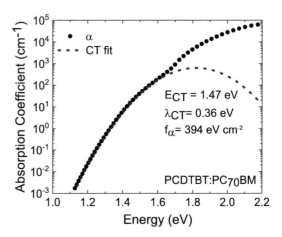

2.6 Aim and Structure of the Book

In Chap. 1 an introduction to organic semiconductor materials and working principles of organic solar cells were presented. In this chapter we had a closer look to the electro-optics of organic solar cells and we presented some of the developed theoretical models that construct our understandings. In Chap. 3 we will present our findings on charge generation and transport in organic solar cells with very small amounts of exciton quencher being low-content donor devices or low-content acceptor devices. We will show, despite having low power conversion efficiency, studying these model systems provides us with very important information and understanding about actual BHJ devices. In Chap. 4 we continue to use the low-content impurities devices, this time in BHJ OSC devices, to investigate the effect of trap states. We will show that mid-gap states are universally present in organic photovoltaics, we will discuss charge generation and recombination *via* these states and explain how they affect the thermodynamic limit of organic photovoltaic devices. Knowing the effect of sub-gap features on the thermodynamic limit of open-circuit voltage in BHJ OSC, in Chap. 5 we present our findings about the relationship of the CT state kinetics and bimolecular recombination by using detailed balance analysis together with the accurate characterisation of CT state and free charge parameters. The experimental methods used for the studies are presented in gray boxes in the order of their appearance in the chapters.

References

1. Hecht, E. (2002). *Optics, 5e*. Pearson Education India.
2. Schwinger, J., DeRaad, L., Jr., Milton, K., & Tsai, W. (1998). *Classical electrodynamics*. Boulder: Westview Press.

3. Born, M., & Wolf, E. (2013). *Principles of optics: Electromagnetic theory of propagation, interference and diffraction of light*. Amsterdam: Elsevier.
4. Griffiths, D. (2005). *Introduction to electrodynamics*. Maryland: American Association of Physics Teachers.
5. Fujiwara, H. (2007). *Spectroscopic ellipsometry: Principles and applications*. New York: Wiley.
6. Pettersson, L., Roman, L., & Inganäs, O. (1999). Modeling photocurrent action spectra of photovoltaic devices based on organic thin films. *Journal of Applied Physics, 86*, 487–496.
7. Armin, A., Velusamy, M., Wolfer, P., Zhang, Y., Burn, P., Meredith, P., & Pivrikas, A. (2014). Quantum efficiency of organic solar cells: Electro-optical cavity considerations. *ACS Photonics, 1*, 173–181.
8. Köhler, A., & Bässler, H. (2015). *Electronic processes in organic semiconductors: An introduction*. New York: Wiley.
9. Alvarado, S., Seidler, P., Lidzey, D., & Bradley, D. (1998). Direct determination of the exciton binding energy of conjugated polymers using a scanning tunneling microscope. *Physical Review Letters, 81*, 1082.
10. Hill, I., Kahn, A., Soos, Z., & Pascal, R., Jr. (2000). Charge-separation energy in films of π-conjugated organic molecules. *Chemical Physics Letters, 327*, 181–188.
11. Rau, U. (2007). Reciprocity relation between photovoltaic quantum efficiency and electroluminescent emission of solar cells. *Physical Review B, 76*, 085303.
12. Tang, C. (1986). Two-layer organic photovoltaic cell. *Applied Physics Letters, 48*, 183–185.
13. Yu, G., Gao, J., Hummelen, J., Wudl, F., & Heeger, A. (1995). Polymer photovoltaic cells: Enhanced efficiencies via a network of internal donor-acceptor heterojunctions. *Science, 270*, 1789–1791.
14. Mikhnenko, O., Blom, P., & Nguyen, T. (2015). Exciton diffusion in organic semiconductors. *Energy & Environmental Science, 8*, 1867–1888.
15. Lin, J., Mikhnenko, O., Chen, J., Masri, Z., Ruseckas, A., Mikhailovsky, A., et al. (2014). Systematic study of exciton diffusion length in organic semiconductors by six experimental methods. *Materials Horizons, 1*, 280–285.
16. Lunt, R., Giebink, N., Belak, A., Benziger, J., & Forrest, S. (2009). Exciton diffusion lengths of organic semiconductor thin films measured by spectrally resolved photoluminescence quenching. *Journal of Applied Physics, 105*, 053711.
17. Shaw, P., Ruseckas, A., & Samuel, I. (2008). Exciton diffusion measurements in poly (3-hexylthiophene). *Advanced Materials, 20*, 3516–3520.
18. Armin, A., Kassal, I., Shaw, P., Hambsch, M., Stolterfoht, M., Lyons, D., et al. (2014). Spectral dependence of the internal quantum efficiency of organic solar cells: Effect of charge generation pathways. *Journal of the American Chemical Society, 136*, 11465–11472.
19. Stoltzfus, D., Donaghey, J., Armin, A., Shaw, P., Burn, P., & Meredith, P. (2016). Charge generation pathways in organic solar cells: Assessing the contribution from the electron acceptor. *Chemical Reviews, 116*, 12920–12955.
20. Marcus, R. (1956). On the theory of oxidation-reduction reactions involving electron transfer. I. *The Journal of Chemical Physics, 24*, 966–978.
21. Hal, P., Meskers, S., & Janssen, R. (2004). Photoinduced energy and electron transfer in oligo(p-phenylene vinylene)-fullerene dyads. *Applied Physics A, 79*, 41–46.
22. Petrella, A., Cremer, J., De Cola, L., Bäuerle, P., & Williams, R. (2005). Charge transfer processes in conjugated Triarylamine-Oligothiophene-Perylenemonoimide dendrimers. *The Journal of Physical Chemistry A, 109*, 11687–11695.
23. Atxabal, A., Arnold, T., Parui, S., Hutsch, S., Zuccatti, E., Llopis, R., Cinchetti, M., Casanova, F., Ortmann, F., & Hueso, L. (2019). Tuning the charge flow between Marcus regimes in an organic thin-film device. *Nature Communications, 10*, 2089.
24. Ward, A., Ruseckas, A., Kareem, M., Ebenhoch, B., Serrano, L., Al-Eid, M., Fitzpatrick, B., Rotello, V., Cooke, G., & Samuel, I. (2015). The impact of driving force on electron transfer rates in photovoltaic donor-acceptor blends. *Advanced Materials, 27*, 2496–2500.
25. Onsager, L. (1938). Initial recombination of ions. *Physical Review, 54*, 554.

26. Braun, C. (1984). Electric field assisted dissociation of charge transfer states as a mechanism of photocarrier production. *The Journal of Chemical Physics, 80*, 4157–4161.
27. Hood, S., Zarrabi, N., Meredith, P., Kassal, I., & Armin, A. (2019). Measuring energetic disorder in organic semiconductors using the photogenerated charge-separation efficiency. *The Journal of Physical Chemistry Letters, 10*, 3863–3870.
28. Deng, W., & Goddard, W. (2004). Predictions of hole mobilities in oligoacene organic semiconductors from quantum mechanical calculations. *The Journal of Physical Chemistry B, 108*, 8614–8621.
29. Li, H., Wang, X., & Li, Z. (2012). Theoretical study of the effects of different substituents of tetrathiafulvalene derivatives on charge transport. *Chinese Science Bulletin, 57*, 4049–4056.
30. Langevin, P. (1903). Recombinaison et mobilites des ions dans les gaz. *Annales de chimie et de physique, 28*, 122.
31. Koster, L., Mihailetchi, V., & Blom, P. (2006). Bimolecular recombination in polymer/fullerene bulk heterojunction solar cells. *Applied Physics Letters, 88*, 052104.
32. Groves, C., & Greenham, N. (2008). Bimolecular recombination in polymer electronic devices. *Physical Review B, 78*, 155205.
33. Heiber, M., Baumbach, C., Dyakonov, V., & Deibel, C. (2015). Encounter-limited charge-carrier recombination in phase-separated organic semiconductor blends. *Physical Review Letters, 114*, 136602.
34. Shoaee, S., Armin, A., Stolterfoht, M., Hosseini, S., Kurpiers, J., & Neher, D. (2019). Decoding charge recombination through charge generation in organic solar cells. *Solar RRL, 3*, 1900184.
35. Shockley, W., & Read, W., Jr. (1952). Statistics of the recombinations of holes and electrons. *Physical Review, 87*, 835.
36. Hall, R. (1959). Recombination processes in semiconductors. *Proceedings of the IEE-Part B: Electronic and Communication Engineering, 106*, 923–931.
37. Würfel, P., & Würfel, U. (2016). *Physics of solar cells: From basic principles to advanced concepts*. New York: Wiley.
38. Sandberg, O., & Armin, A. (2019). On the effect of surface recombination in thin film solar cells, light emitting diodes and photodetectors. *Synthetic Metals, 254*, 114–121.
39. Shockley, W., & Queisser, H. (1961). Detailed balance limit of efficiency of p-n junction solar cells. *Journal of Applied Physics, 32*, 510–519.
40. Vandewal, K., Albrecht, S., Hoke, E., Graham, K., Widmer, J., Douglas, J., Schubert, M., Mateker, W., Bloking, J., Burkhard, G., Sellinger, A., Fréchet, J., Amassian, A., Riede, M., McGehee, M., Neher, D., & Salleo, A. (2014). Efficient charge generation by relaxed charge-transfer states at organic interfaces. *Nature Materials, 13*, 63–68.
41. Vandewal, K., Tvingstedt, K., Gadisa, A., Inganäs, O., & Manca, J. (2009). On the origin of the open-circuit voltage of polymer-fullerene solar cells. *Nature Materials, 8*, 904–909.
42. Taylor, N., & Kassal, I. (2018). Generalised Marcus theory for multi-molecular delocalised charge transfer. *Chemical Science, 9*, 2942–2951.
43. Kaiser, C., Zeiske, S., Meredith, P., & Armin, A. (2020). Determining ultralow absorption coefficients of organic semiconductors from the sub-bandgap photovoltaic external quantum efficiency. *Advanced Optical Materials, 8*, 1901542.

Chapter 3
Anomalous Exciton Quenching in Organic Semiconductors

Abstract The dynamics of exciton quenching are critical to the operational performance of organic optoelectronic devices, but their measurement and elucidation remains an ongoing challenge. The work presented in this chapter describes a method for quantifying small photoluminescence quenching efficiencies of organic semiconductors under steady state conditions. Exciton quenching efficiencies of three exemplary organic semiconductors, $PC_{70}BM$, P3HT and, PCDTBT are measured at different bulk quencher densities under continuous small irradiance ($\sim\mu W\ cm^{-2}$). By implementing a steady state bulk-quenching model, exciton diffusion lengths for the studied materials are determined. At low quencher densities we find that a secondary quenching mechanism is in effect which is responsible for approximately 20% of the total quenched excitons. This quenching mechanism is observed in all three studied materials and exhibits quenching volumes on the order of several thousand nm^3. The exact origin of this quenching process is not clear, but it may be indicative of delocalised excitons being quenched prior to thermalisation. This chapter is written based on a collaborative work published by the author in the Journal of Physical Chemistry Letters (JPCL) with permission from Ref. [1] copyright(2018) American Chemical Society.

3.1 Introduction

As explained in Chaps. 1 and 2, organic semiconductors are generally disordered materials with low permittivities and strongly bound photo-excitations (excitons) at room temperature [2, 3]. Exciton migration and the associated dynamics play important roles in defining the performance of organic optoelectronic devices, including organic solar cells (OSC), organic light emitting diodes (OELDs), organic photodetectors, and sensors operating based on exciton quenching [4–7]. Exciton migration through organic semiconductors is diffusive, described by site-to-site hopping of localised excitons. Significant effort has been expended to evaluate the diffusion

© The Author(s), under exclusive license to Springer Nature Switzerland AG 2022

N. Zarrabi, *Optoelectronic Properties of Organic Semiconductors*,

SpringerBriefs in Materials,

https://doi.org/10.1007/978-3-030-93162-9_3

lengths of singlet excitons in particular since they are the more prevalent species [8–12]. In this regard, exciton diffusion lengths are often evaluated by studying the dynamics of exciton quenching in the presence of quenchers; specifically, time-resolved photoluminescence (PL) measurements can be used to evaluate exciton life-times and diffusion coefficients, providing sufficient information to infer the exciton diffusion lengths [13–18].

However, the signal-to-noise-ratio (SNR) of PL quenching measurements approaches zero at low quenching efficiencies due to the background emission act-ing as noise, which contains sample-to-sample variation and photon shot noise in low-emissive yield systems. It has therefore been very challenging to quantify small exciton quenching yields, a problem addressed by Siegmund et al. [19] using 1-D modelling of solar cell photocurrent spectra to extract exciton diffusion lengths even in non-fluorescent materials.

To investigate exciton quenching at low quencher densities (i.e. in the low yield limit), we developed a method for measuring exciton quenching efficiencies under steady state conditions using low-irradiance thermal light. The technique relies upon a steady state 3-D quenching model that can fit to experimental results to directly quantify exciton diffusion lengths, with no requirement for knowledge of the exciton lifetimes and diffusion coefficients. Importantly, because this method is background-free, i.e., it is only sensitive to the quenched part of the PL signal—it remains accurate at low quenching efficiencies. In this regime, we observe an anomalous exciton quenching pathway that is absent at high yields and hence, not previously observed in transient measurements. This secondary quenching pathway corresponds to large quenching volumes and may originate from the quenching of delocalised excitons prior to thermalisation [20–23].

3.2   Theoretical Framework

The quenching efficiency can be related to the quencher density by adopting a sta-tistical probability approach as follows: Consider a molecular semiconductor matrix slightly doped with an exciton quenching material, so that the ith quencher molecule is located at position $\mathbf{r}_i$. The probability that an exciton initially at $\mathbf{r}$ will be quenched at quencher i as $p(|\mathbf{r} - \mathbf{r}_i|)$, for a monotonically decreasing function $p(r)$. The prob-ability that the exciton is quenched by any of the N quenchers within the matrix is then $1 - \prod_{i=1}^{N}[1 - p(|\mathbf{r} - \mathbf{r}_i|)]$. The observable quenching yield (QY) is obtained by averaging the quenching probability over the initial position of the exciton:

$$QY = \frac{1}{V} \int \int \int \left[1 - \prod_{i=1}^{N}[1 - p(|\mathbf{r} - \mathbf{r}_i|)] \right] d\mathbf{r}. \qquad (3.1)$$

The QY increases linearly with the quencher density at low densities, where the quenching volumes of individual quenchers do not overlap. It deviates from linearity

Fig. 3.1 Exciton quenching yield *versus* quencher number density under steady state conditions. Each quencher is surrounded by a sphere indicating its quenching volume. At low densities, increasing the number of quenchers results in a linear increase in the quenching yield. A deviation from linearity occurs as the quenching volumes start overlapping. Ultimately, the whole space is quenched, and a saturation is achieved

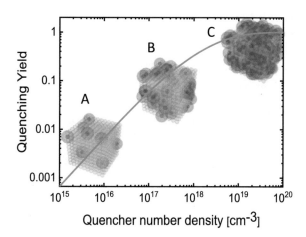

with increasing quencher density until saturation is achieved, where the whole space is covered by the quenching volumes (Fig. 3.1).

To compute the QY, we first determine $p(r)$ by assuming a quencher of radius a centred at the origin of a 1-D lattice with lattice constant δ. The exciton undergoes a random walk with lifetime τ_D and is quenched if found at the position $x = a$. The probability $p(x)$ obeys the following relation:

$$p(x) = \frac{1}{2}[p(x - \delta) + p(x + \delta)] - \frac{\Delta t}{\tau_D} p(x), \tag{3.2}$$

where Δt is the time of each jump. The first, lossless term indicates that the survival probability for a walker starting at x equals the average of the probabilities for a walker starting from the points reachable in one step from x. The second term adds loss, i.e., some probability is lost during each jump to ensure an exponential decay with lifetime τ_D. From the definition of second derivative, in the continuum limit ($\delta \to 0$) Eq. 3.2 becomes:

$$\frac{d^2}{dx^2} p(x) - \frac{2\Delta t}{\delta^2 \tau_D} p(x) = 0, \tag{3.3}$$

where $\frac{\delta^2}{2\Delta t}$ equals the diffusion constant D (in 1-D). In 3-D, this becomes the spherically symmetric Helmholtz equation:

$$\nabla^2 p(\mathbf{r}) - \frac{1}{D\tau_D} p(\mathbf{r}) = 0, \tag{3.4}$$

whose radial part is solved by a spherical Bessel functions of order zero. In particular, the real solution obeying the boundary conditions $p(a) = 1$ and $p(\infty) = 0$ is a spherical Hankel function of the second kind:

Fig. 3.2 The predicted exciton quenching yield plotted *versus* the number density of quenchers for exciton diffusion lengths between 2 and 18 nm. Saturation occurs at lower concentrations for larger exciton diffusion lengths as one would intuitively expect

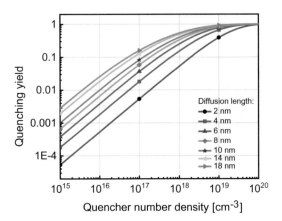

$$p(r) = \frac{a}{r} \exp\left(-\frac{r - a}{\sqrt{D\tau_{\mathrm{D}}}}\right) \tag{3.5}$$

where $l_{\mathrm{diff}} = \sqrt{D\tau_{\mathrm{D}}}$ is 1-D diffusion length [24]. Substituting Eq. 3.5 into Eq. 3.1, we can compute the QY for different exciton diffusion lengths as a function of quencher number density (Fig. 3.2). The quencher radius is considered to be 0.75 nm, which is a typical dimension for organic semiconductors.

The experimental parameter that allows evaluation of small quenching yields is the internal quantum efficiency (IQE) of the solar cells containing the three 'neat' semiconductors with varying amounts of quencher. The IQE is the product of exciton quenching yield, charge transfer (η_{CT}) and charge collection (η_{CC}) efficiencies [25]:

$$\mathrm{IQE} = \mathrm{QY} \times \eta_{\mathrm{CT}} \times \eta_{\mathrm{CC}}, \tag{3.6}$$

The charge collection and transfer efficiencies in the devices with very low amount of quencher (<1%) are invariant with the density of quencher as the cells operate far below the charge transport percolation threshold of the quenching molecules. In other words, the distance between two donors (or acceptors) in a low-donor-(or acceptor)-content device is larger than the average charge hopping distance for a hole(electron). In a relevant work [26] we have shown that donor molecules in a low-donor-content device behave as trap sites in the acceptor energy gap. The kinetics of the trapped holes can be describe by the SRH statistics where hole transport can be explained by the process of optical release.

A schematic energy level diagram of a low content donor device and the procedure of hole transport is shown in Fig. 3.3 in subsequent steps. Photogenerated excitons, are induced by photon absorption in the acceptor. The excitons are predominantly quenched and dissociate at donor sites followed by a hole transfer from acceptor to the donor. This results in the generation of a free electron in the acceptor and a (trapped) hole in the donor. In the next step (absorption of the next photon), a photoinduced electron transfer from the acceptor exciton to the positively charge donor site will

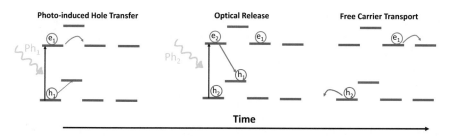

Fig. 3.3 Schematic energy level diagram of an acceptor (in blue) matrix containing trap-like donor (in red) sites and the procedure of optical release is shown

happen. This produces a free hole in the HOMO level of the acceptor and neutralizes the donor site. As a result, the transport of free electrons and holes then exclusively takes place in the acceptor, whereas the donor sites act as traps for holes. This can explain the unusually large open circuit voltage in the so-called low donor content solar cells [27–29]. According to this new picture, the V_{OC} in these devices is larger than that predicted by CT state energetics and it is defined by the energy gap of the host material. By increasing the amount of donor to values exceeding 1%, donor to donor hopping might become possible and the open circuit voltage reduced and dominated by the donor-acceptor CT states. A similar statement holds for electron transport in the low acceptor devices.

Consequently, in the very low quencher limit the IQE of the devices only depend on the QY and the charge transfer efficiency.

3.3 Experimental Results

We experimentally studied two polymeric semiconductors, P3HT and PCDTBT and a fullerene derivative $PC_{70}BM$ for analysis with this model. In this regard, standard architecture organic solar cells with a structure ITO/PEDOT:PSS/semiconductor/Al were fabricated. Notably, semiconducting active layers were prepared from solutions at different concentrations of the exciton quenching material ($PC_{70}BM$ is the quencher in the polymers matrices and TAPC is the quencher in $PC_{70}BM$ matrix) via sequential dilution. The details of device fabrication can be found in Appendix.

In Fig. 3.4 the representative external quantum efficiencies (EQE_{PV}) of the 'low-donor-content' ($TAPC:PC_{70}BM$) and low-acceptor-content ($PCDTBT:PC_{70}BM$ and $P3HT:PC_{70}BM$) devices measured at short circuit for different densities of exciton quencher are shown in panel "a" for each material system. By considering the parasitic absorptions and interference effects in the solar cell stack via a transfer matrix analysis, it is possible to accurately determine the IQE from this measured EQE_{PV} (See Box 2.2). This is shown in panel "b" of Fig. 3.3 for the three material systems [30]. The extinction coefficients and refractive indices of the materials of the stack are presented in Fig. 3.4. In order to quantify the charge generation mediated only by the quencher molecules, we subtracted the IQE of the neat semiconductor device ($PC_{70}BM$-only junction for low-donor-content device and PCDTBT and P3HT-only junction for low-acceptor-content device) from all the other cells with different quencher concentrations $\mathbf{c}$.

$$IQE_{DA}(\mathbf{c}) = IQE_{DA}(\mathbf{c}) - IQE_{DA}(\mathbf{c} = 0) \qquad (3.7)$$

This approach delivers the quencher-induced-IQE, $IQE_{DA}(\mathbf{c})$ in which the contribution of excitons quenched by any means other than via the quencher molecules is excluded. These contribution may arise (for example) from self-quenching within the disordered density of states of the semiconductor matrix or through trap states [31]. Quencher-induced-IQE spectra are shown in panels "c" of Fig. 3.3 for each material system. They are incident illumination energy independent within experimental uncertainty. This is expected for charge generation at a donor-acceptor interface while self-quenching within the neat material matrix IQE ($\mathbf{c} = 0$) is excitation energy dependent.

The ultimate results of this analysis are shown in Fig. 3.5 for all three systems studied ($PC_{70}BM$, PCDTBT and P3HT) in which the exciton quenching yield is plotted *versus* the number density of quenchers (calculated from the weight ratios and densities). At low quencher concentrations (<1 wt%) quenching yields were simply determined from the quencher-induced-IQEs of Fig. 3.3 (panels c) (open symbols in Fig. 3.7). At higher quencher concentrations (>1 wt% and into the saturation regime shown in Fig. 3.1) the QY was directly measured from steady state PL measurements on films without the ancillary solar cell layers (filled symbols in Fig. 3.7) (Fig. 3.6).

The quencher-induced-IQE values were then normalized to match the PL data at shared quencher density data points for self-consistency. Experimental limitations precluded PL quenching measurements on the $PC_{70}BM$ system. However, fortunately the saturation of the QY in the fullerene was almost reached using the quencher-induced-IQEs to satisfy the model fitting. We should note that because at low quencher concentrations our method uses charge carrier photogeneration as a probe for exciton quenching, it is insensitive to the long-range energy transfer from a molecule to the quencher. This mechanism has been found to play role in P3HT:fullerene blends [32]. It is clear from the data in Fig. 3.5 that the quenching yields behave differently to that predicted for the diffusion-only case (Fig. 3.2). Typical diffusion lengths from 4 to 8 nm can fit the higher quencher densities $>10^{18}$ cm^{-3}. However, at lower concentrations, there is a clear anomalous trend. The exciton quenching yields are far larger than expected from an extrapolation of the diffusion-regime at low quencher densities. This secondary pathway, however, shows a linear increase at the lowest densities which saturates at mid quencher densities—this indicates the pathway is not particularly efficient. Since the exact mechanism for this quenching pathway is unclear, we cannot provide a probability function for quenching through this process that could be used in Eq. 3.1. As such we used a simple and generic model suggested by Perrin [33] that yields the quenching volume:

$$QY = 1 - \exp(-\mathbf{c}V), \qquad (3.8)$$

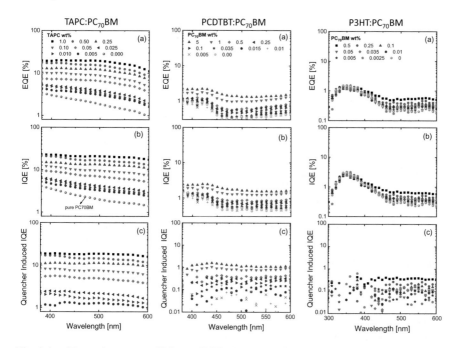

Fig. 3.4 a External quantum efficiency (EQE_{PV}) spectra of devices containing $PC_{70}BM$ as the matrix material and different concentrations of TAPC (wt%) as the quencher and PCDTBT and P3HT as the matrix and $PC_{70}BM$ (wt%) as the quencher. **b** Internal quantum efficiency (IQE) evaluated for each device from the EQE_{PV} with analysis of parasitic absorption and interference effects. At low quencher concentrations, a wavelength-dependent IQE is observed due to wavelength-dependent (illumination energy dependent) charge generation in predominantly neat material matrix. **c** Quencher-induced-IQE spectra for charge generation via donor:acceptor pairs evaluated by subtracting the IQE of the neat material device. These IQEs show no significant wavelength dependence

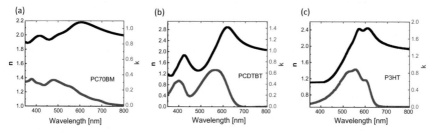

Fig. 3.5 Optical constants: Extinction coefficient **k** and index of refraction **n** for **a** $PC_{70}BM$ **b** PCDTBT, and **c** P3HT

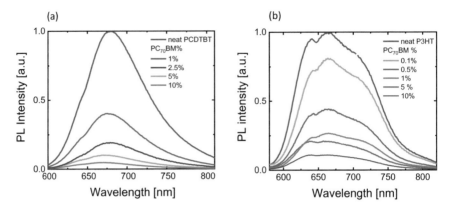

Fig. 3.6 The photoluminescence (PL) spectra of P3HT:PC$_{70}$BM (**a**) and PCDTBT: PC$_{70}$BM (**b**), measured with PC$_{70}$BM concentrations as shown. The PL quenching yield (PLQY) was calculated by integrating the PL intensity: PLQY $= \frac{\int_{\lambda_1}^{\lambda_2} I_{neat}d\lambda - \int_{\lambda_1}^{\lambda_2} I_{doped}d\lambda}{\int_{\lambda_1}^{\lambda_2} I_{neat}d\lambda}$

where, **c** is the quencher concentration and V corresponds to the "quenching volume". The total quenching yield can be then written as the superposition of the diffusion and the anomalous quenching yields:

$$QY_{total} = \gamma_{an}QY_{an} + (1 - \gamma_{an})QY_{diff}, \qquad (3.9)$$

in which γ_{an} represents the contribution of the efficiency of the anomalous quenching pathway. The final results of the fitting based on Eq. 3.9 are shown in Fig. 3.5 (solid lines) yielding the diffusion length and quenching volume for each material system.

The values of the exciton diffusion lengths for PC$_{70}$BM, PCDTBT, and P3HT are comparable with previous reports using time resolved photoluminescence [8, 16, 17].The anomalous quenching pathways plays a significant role in the quenching at low quencher densities with large quenching volumes of 9325 nm^3, 14900 nm^3 and 2381 nm^3 in PC$_{70}$BM, P3HT and PCDTBT respectively. It is plausible that such large quenching volumes may be due to delocalized exciton density formed at very early times of photoexcitation within the disordered landscape as shown recently by Mannouch et al. [34]. This initially delocalized exciton can be quenched at a finite distance from the quencher prior to density localization. This pathway is not efficient and does not play a substantial role in the high quenching density regime. However, further understanding this mechanism may result in better material optimization for optoelectronic devices requiring exciton migration. The observation of a pathway with large quenching volumes is also an intriguing fundamental insight into disordered semiconductors requiring additional careful analysis.

Fig. 3.7 Exciton quenching efficiency as a function of quencher number density in PC$_{70}$BM, P3HT, and PCDTBT. In all three materials, the data can be described using a steady state exciton diffusion model at high quencher densities (dotted lines), while anomalously strong quenching is observed at low quencher concentrations (dashed lines). The solid line is a fit to Eq. 3.9, yielding exciton diffusion lengths and the quenching volume for the anomalous quenching mechanism. The vertical error bars correspond to one standard deviation of the quencher induced-IQEs and the horizontal error bars correspond to the concentration uncertainty calculated for sequential dilutions

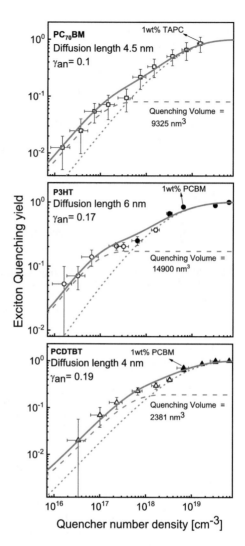

3.4 Conclusion

In conclusion, we have presented a new way of measuring small exciton quenching yields based on charge photogeneration (IQE) measurements in the steady state. A 3-D steady state quenching model has been developed and applied to the experimental results of quenching efficiency *versus* quencher density for three organic semiconductors. The exciton diffusion lengths extracted from the higher quencher density regime are consistent with previous reports using standard PL quenching methods. The improved signal-to-noise ratio of our technique is a key feature and allows for quantification of the quenching yields at low quencher densities. In this regime, we found an anomalous quenching pathway which is not efficient but long range. This may be related to a question that has been a matter of some debate in the organic semiconductor community, namely: whether exciton quenching occurs at the interface between the matrix and the quencher (localised excitons) or remotely at a certain distance from it through the existence of a certain delocalisation especially under incoherent illumination such as the thermal light. As such these observations may be indicative of delocalised excitons being quenched at a certain distance from the quencher before they thermalise.

References

1. Zarrabi, N., Yazmaciyan, A., Meredith, P., Kassal, I., & Armin, A. (2018). Anomalous exciton quenching in organic semiconductors in the low-yield limit. *The Journal of Physical Chemistry Letters, 9,* 6144–6148.
2. Brédas, J., Norton, J., Cornil, J., & Coropceanu, V. (2009). Molecular understanding of organic solar cells: The challenges. *Accounts of Chemical Research, 42,* 1691–1699.
3. Feron, K., Zhou, X., Belcher, W., & Dastoor, P. (2012). Exciton transport in organic semiconductors: Förster resonance energy transfer compared with a simple random walk. *Journal of Applied Physics, 111,* 044510.
4. Menke, S., & Holmes, R. (2014). Exciton diffusion in organic photovoltaic cells. *Energy & Environmental Science, 7,* 499–512.
5. Tamai, Y., Ohkita, H., Benten, H., & Ito, S. (2015). Exciton diffusion in conjugated polymers: From fundamental understanding to improvement in photovoltaic conversion efficiency. *The Journal of Physical Chemistry Letters, 6,* 3417–3428.
6. Reineke, S., & Baldo, M. (2012). Recent progress in the understanding of exciton dynamics within phosphorescent OLEDs. *Physica Status Solidi (a), 209,* 2341–2353.
7. Arquer, F., Armin, A., Meredith, P., & Sargent, E. (2017). Solution-processed semiconductors for next-generation photodetectors. *Nature Reviews Materials, 2,* 1–17.
8. Mikhnenko, O., Blom, P., & Nguyen, T. (2015). Exciton diffusion in organic semiconductors. *Energy & Environmental Science, 8,* 1867–1888.
9. Lin, J., Mikhnenko, O., Chen, J., Masri, Z., Ruseckas, A., Mikhailovsky, A., et al. (2014). Systematic study of exciton diffusion length in organic semiconductors by six experimental methods. *Materials Horizons, 1,* 280–285.
10. Lunt, R., Giebink, N., Belak, A., Benziger, J., & Forrest, S. (2009). Exciton diffusion lengths of organic semiconductor thin films measured by spectrally resolved photoluminescence quenching. *Journal of Applied Physics, 105,* 053711.
11. Leow, C., Ohnishi, T., & Matsumura, M. (2014). Diffusion lengths of excitons in polymers in relation to external quantum efficiency of the photocurrent of solar cells. *The Journal of Physical Chemistry C, 118,* 71–76.
12. Luhman, W., & Holmes, R. (2011). Investigation of energy transfer in organic photovoltaic cells and impact on exciton diffusion length measurements. *Advanced Functional Materials, 21,* 764–771.
13. Markov, D., Amsterdam, E., Blom, P., Sieval, A., & Hummelen, J. (2005). Accurate measurement of the exciton diffusion length in a conjugated polymer using a heterostructure with a side-chain cross-linked fullerene layer. *The Journal of Physical Chemistry A, 109,* 5266–5274.
14. Markov, D., Hummelen, J., Blom, P., & Sieval, A. (2005). Dynamics of exciton diffusion in poly(p-phenylene vinylene)/fullerene heterostructures. *Physical Review B, 72,* 045216.
15. Scully, S., & McGehee, M. (2006). Effects of optical interference and energy transfer on exciton diffusion length measurements in organic semiconductors. *Journal of Applied Physics, 100,* 034907.

16. Mikhnenko, O., Azimi, H., Scharber, M., Morana, M., Blom, P., & Loi, M. (2012). Exciton diffusion length in narrow bandgap polymers. *Energy & Environmental Science, 5,* 6960.
17. Shaw, P., Ruseckas, A., & Samuel, I. (2008). Exciton diffusion measurements in Poly(3-hexylthiophene). *Advanced Materials, 20,* 3516–3520.
18. Hedley, G., Ward, A., Alekseev, A., Howells, C., Martins, E., Serrano, L., et al. (2013). Determining the optimum morphology in high-performance polymer-fullerene organic photovoltaic cells. *Nature Communications, 4,* 2867.
19. Siegmund, B., Sajjad, M., Widmer, J., Ray, D., Koerner, C., Riede, M., et al. (2017). Exciton diffusion length and charge extraction yield in organic bilayer solar cells. *Advanced Materials (Deerfield Beach, Fla.), 29,* 1–5.
20. Gelinas, S., Rao, A., Kumar, A., Smith, S., Chin, A., Clark, J., et al. (2014). Ultrafast long-range charge separation in organic semiconductor photovoltaic diodes. *Science, 343,* 512–516.
21. Piris, J., Dykstra, T., Bakulin, A., Loosdrecht, P., Knulst, W., Trinh, M., et al. (2009). Photogeneration and Ultrafast Dynamics of Excitons and Charges in P3HT/PCBM Blends. *The Journal of Physical Chemistry C, 113,* 14500–14506.
22. Wang, H., Wang, H., Gao, B., Wang, L., Yang, Z., Du, X., et al. (2011). Exciton diffusion and charge transfer dynamics in nano phase-separated P3HT/PCBM blend films. *Nanoscale, 3,* 2280.
23. Kaake, L., Jasieniak, J., Bakus, R., Welch, G., Moses, D., Bazan, G., & Heeger, A. (2012). Photoinduced charge generation in a molecular bulk heterojunction material. *Journal of the American Chemical Society, 134,* 19828–19838.
24. Berg, H. (1993). *Random walks in biology*. Princeton: Princeton University Press.
25. Rand, B., Burk, D., & Forrest, S. (2007). Offset energies at organic semiconductor heterojunctions and their influence on the open-circuit voltage of thin-film solar cells. *Physical Review B, 75,* 115327.
26. Sandberg, O., Zeiske, S., Zarrabi, N., Meredith, P., & Armin, A. (2020). Charge carrier transport and generation via trap-mediated optical release in organic semiconductor devices. *Physical Review Letters, 124,* 128001.
27. Zhang, M., Wang, H., Tian, H., Geng, Y., & Tang, C. (2011). Bulk heterojunction photovoltaic cells with low donor concentration. *Advanced Materials, 23,* 4960–4964.
28. Vandewal, K., Widmer, J., Heumueller, T., Brabec, C., McGehee, M., Leo, K., et al. (2014). Increased open-circuit voltage of organic solar cells by reduced donor-acceptor interface area. *Advanced Materials, 26,* 3839–3843.
29. Collado-Fregoso, E., Pugliese, S., Wojcik, M., Benduhn, J., Bar-Or, E., Perdigón Toro, L., Hörmann, U., Spoltore, D., Vandewal, K., Hodgkiss, J. & Neher, D. (2019). Energy-gap law for photocurrent generation in fullerene-based organic solar cells: the case of low-donor-content blends. *Journal of the American Chemical Society, 141,* 2329–2341.
30. Armin, A., Velusamy, M., Wolfer, P., Zhang, Y., Burn, P., Meredith, P., & Pivrikas, A. (2014). Quantum efficiency of organic solar cells: Electro-optical cavity considerations. *ACS Photonics, 1,* 173–181.
31. Mikhnenko, O., Kuik, M., Lin, J., Kaap, N., Nguyen, T., & Blom, P. (2014). Trap-limited exciton diffusion in organic semiconductors. *Advanced Materials, 26,* 1912–1917.
32. Long, Y., Ward, A., Ruseckas, A., & Samuel, I. (2016). Effect of a high boiling point additive on the morphology of solution-processed P3HT-fullerene blends. *Synthetic Metals, 216,* 23–30.
33. Perrin, F. (1924). Loi de decroissance du pouvoir fluorescent en fonction de la concentration. *Comptes Rendus de l'Academie des Sciences, 178,* 1978–1980.
34. Mannouch, J., Barford, W., & Al-Assam, S. (2018). Ultra-fast relaxation, decoherence, and localization of photoexcited states in -conjugated polymers. *The Journal of Chemical Physics, 148,* 034901.

Chapter 4
On the Effect of Mid-Gap Trap States on the Thermodynamic Limit of OPV Devices

Abstract Detailed balance is a cornerstone of our understanding of artificial light-harvesting systems. For next generation organic solar cells, this involves intermolecular charge-transfer (CT) states whose energies set the maximum open circuit voltage V_{OC}. In the study presented in this chapter, sub-gap states significantly lower in energy than the CT states in the external quantum efficiency spectra of a substantial number of organic semiconductor blends have directly been observed. Taking these states into account and using the principle of reciprocity between emission and absorption results in non-physical radiative limits for the V_{OC}. Compelling evidence has been provided for these states being non-equilibrium mid-gap traps which contribute to photocurrent by a non-linear process of optical release, upconverting them to the CT state. This motivates the implementation of a two-diode model which is often used in emissive inorganic semiconductors. The model accurately describes the dark current, V_{OC} and the long-debated ideality factor in organic solar cells. Additionally, the charge-generating mid-gap traps have important consequences for our current understanding of both solar cells and photodiodes—in the latter case defining a detectivity limit several orders of magnitude lower than previously thought. This chapter is written based on collaborative work published by the author in Nature Communications in 2020 [1].

4.1 Introduction

The basic thermodynamic principle of detailed balance is fundamental in defining the maximum efficiency with which a semiconductor with a certain bandgap can convert photons to electrical power via the photovoltaic effect. In particular, detailed balance provides a means to predict the theoretical limit of the open circuit voltage, V_{OC} and short circuit current, J_{Ph} of any solar cell [2]. A further consequence of detailed balance is the so called reciprocity relation between the photovoltaic external quantum efficiency (EQE_{PV}) and the electroluminescence quantum efficiency (EQE_{LED}),

© The Author(s), under exclusive license to Springer Nature Switzerland AG 2022   49
N. Zarrabi, *Optoelectronic Properties of Organic Semiconductors*,
SpringerBriefs in Materials,
https://doi.org/10.1007/978-3-030-93162-9_4

i.e. the relative efficiency with which any particular device turns light into electrical current and vice versa, current into light. Stated simply, a good solar cell should in principle be a good light emitting diode (LED). The reciprocity relation further enables us to derive, respectively, the radiative limit of the open circuit voltage (V_{OC}^{Rad}) and the non-radiative loss of the open circuit voltage ΔV_{OC}^{NR} [3]. This information can then be used to obtain a more realistic calculation of V_{OC} which is usually close (ideally identical) to the experimentally measured value [4–7]. Photovoltaics as a field has relied upon detailed balance and reciprocity since its inception in the early 1960s [2]—irrespective of the semiconductor in question be it c-Si, GaAs or more latterly lead halide perovskites and organics.

In accordance with the reciprocity principle, the V_{OC} of a solar cell is generally regarded to be ultimately limited by the lowest-lying charge-generating energy states to which the system thermalises and finds a (quasi) equilibrium condition. In organic semiconductors, these states are thought to be the CT states at the donor-acceptor interface [8–10]. The CT states are sub-gap with an additional consequential voltage loss relative to the lowest energy singlet exciton state. Being sub-gap, CT states have weak oscillator strength—they do not absorb strongly (nor emit) and thus have very low EQE_{PV}. Depending on their energy offset relative to the singlet excitons, they can also be very difficult to identify—this is a particular emerging problem in the so-called non-fullerene acceptor (NFA) systems which are delivering record power conversion efficiencies of 18% [11] and have low or negligible offsets [12, 13]. These materials are challenging our long-held views on the dynamics of (or indeed the need for) the CT state and therefore the nature of detailed balance and reciprocity in organic BHJ solar cells.

Given the above energetic considerations, it is thus clear that accurate determination of CT states is inevitably limited by the accuracy with which EQE_{PV} can be measured. Indeed, this is a generic issue in studying sub-gap features across all semiconductors. Motivated by the emerging reciprocity question in organic photovoltaics and this broader sub-gap issue, [14–16] ultra-sensitive photocurrent measurements with detection limits within a fraction of fA have been presented. This allows EQE_{PV}s as low as 10^{-10} to be reliably determined at wavelengths up to 2400 nm [17]. Ultra-sensitive EQE_{PV}s are presented for organic and inorganic semiconductor solar cells including a number of the recently introduced NFA systems. Notably, distinct sub-gap features in a large variety of organic semiconductor blends at energies well below the CT state have been observed. Including these additional low energy states in the calculation of V_{OC}^{Rad} from EQE_{PV} (as one would using the principle of reciprocity) results in considerably lower apparent non-radiative losses than determined from EQE_{LED}. This appears to contradict reciprocity between absorption and emission which is valid for systems in thermodynamic equilibrium.

In the study presented in this chapter, these observations have been rationalised by providing compelling evidence that the low energy absorptions arise from partially radiative mid-gap trap states. These states can contribute to photocurrent generation by optical release which upconverts the non-equilibrium traps to the CT state energy but also give rise to radiative emission of photons with energies well below the gap. These non-linear processes explain the apparent violation from the equilibrium detailed balance but demands a modified picture for organic solar cells (and indeed photodiodes) to incorporate the non-equilibrium mid-gap trap states.

4.2 Ultra-Sensitive EQE$_{PV}$ Measurements and the Failure of Reciprocity

To date, most sensitive EQE$_{PV}$ measurements have only been able to partly detect the contributions of CT states within the sub-gap region corresponding to signals down to $\sim 10^{-6}$ [18–23]. As a natural consequence, V_{OC} has always been correlated with the CT state energy. However, our measurement setup is able to detect EQE$_{PV}$ signals as low as 10^{-10} and with a spectral window extended to 2400 nm (see Box 4.1). To our knowledge, these are the most sensitive (termed as ultra-sensitive) EQE$_{PV}$ measurements reported thus far in any photovoltaic system.

Box 4.1

Ultra-Sensitive EQE$_{PV}$ Measurement: For this measurement a home-built setup, developed by Zeiske et al. [17], was used. In the setup, a high-performance commercial spectrophotometer with integrated double holographic grating monochromators (PerkinElmer, Lambda950) was used as the light source (LS) providing an extended wavelength regime from 175 nm up to 3300 nm. A multi-blade optical chopper (OC) wheel (Thorlabs, MC2000B) physically chops the probe light at $\omega = 273$ Hz. The chopped light beam then is focused on the device under test (DUT) by using optics. The electric photocurrent of the device first is amplified by passing through a low noise pre-current amplifier with variable gain (Femto, DLPCA-200) and then detected with a lock-in amplifier (Stanford Research Systems, SR860) providing various integration times (electrical bandwidths).

As explained extensively in Ref. [17] in order to achieve high dynamic range and increase the sensitivity of the measurement several consideration has been taken into account to reduce optical and electrical noises in the measurement set-up. The frequency of the measurement (273 Hz) was chosen to be different from the unavoidable flicker noise at low frequencies and the main hum noise at 50 Hz of the measurement setup. To filter out remaining parasitic stray light (optical noise) different OD4 long-pass filters (LPF) were used. Ultimately, the minimum measurable EQE$_{PV,min}$ of a device is determined by the device noise current $I_{noise} = I_{NSD}\sqrt{\Delta f}$ where I_{NSD} is the noise spectral density and $\sqrt{\Delta f}$ is the electrical bandwidth. While the spectral density of the noise is indeed dependent on the shunt resistance, the total noise is also dependent on the electrical bandwidth of the measurement (inversely proportional to the lock-in amplifier time constant). For smaller EQE$_{PV,min}$ to be detected or where the shunt resistance is low, a smaller electrical bandwidth is required. We dynamically varied the electrical bandwidth during the wavelength sweep and truncated the data at the point where signal-to-noise-ratio (SNR), approaches unity. At most wavelength ranges the SNR is greater than 20 dB, up to 90 dB. We note that each spectra may take up to 3 days to complete in solar cells with smaller shunt resistances.

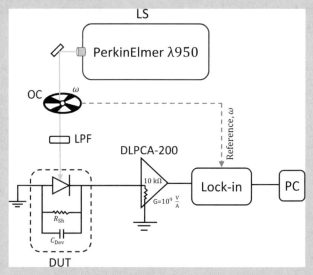

The figure bellow, shows two exemplary ultra-sensitive EQE_{PV} spectra measured with different electrical bandwidths. The associated thermal noise shown as a horizontal line.

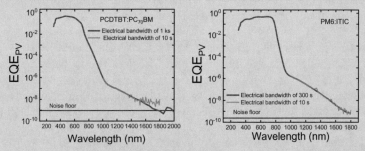

For the calibration process, a Newport NIST-calibrated silicon (818-UV), germanium (818-IR) and Thorlabs indium gallium arsenide (S148C) photodiode sensor were used.

In Fig. 4.1 the measured ultra-sensitive EQE_{PV} for a solar cell devise comprising PM6:ITIC is shown. For comparison, on the right axis of the EQE_{PV}, ϕ_{BB} is plotted *versus* photon energy in order to show the spectral overlap in the sub-gap region. This measurement clearly reveals a sub-gap feature, far below the CT states.

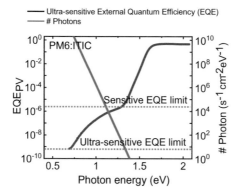

Fig. 4.1 The black curve represents the experimental EQE$_{PV}$ *versus* photon energy for a PM6:ITIC solar cell. The limit of the sensitive EQE$_{PV}$ (reported before) and the ultra-sensitive EQE$_{PV}$ (reported in this work) are shown with dotted lines. The corresponding ϕ_{BB} *versus* photon energy is plotted on the right axis (blue curve). Ultra-sensitive EQE$_{PV}$ measurements reveal a sub-gap feature in the EQE$_{PV}$ spectrum

Fig. 4.2 The calculated $J_0^{Rad} = q \int_{\epsilon_{min}}^{\infty} EQE_{PV}(\epsilon) \times \phi_{BB}(\epsilon)d\epsilon$ (the green curve on the left axis) and $J_{Ph} = q \int_{\epsilon_{min}}^{\infty} EQE_{PV}(\epsilon) \times \phi_{Sun}(\epsilon)d\epsilon$ (the pink curve on the right axis) are shown *versus* the photon energy. For comparison, the CT state energy (green) and optical gap (pink) have been included as indicated by the vertical dashed lines

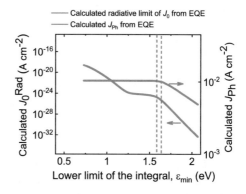

Using the ultra-sensitive EQE$_{PV}$, J_0^{Rad} and J_{Ph} can be determined (see Eq. (2.25), see Eq. (2.26)). Note again that the lower limit ϵ_{min} and the corresponding truncation of J_0^{Rad} and J_{Ph} will have a significant impact on the determined radiative limit. This is demonstrated in Fig. 4.2 where the calculated $J_0^{Rad}(\epsilon_{min})$ and $J_{Ph}(\epsilon_{min})$ *versus* the photon energy are shown.

The theoretical radiative limit (V_{OC}^{Rad}) of V_{OC} can be then determined from J_0^{Rad} and J_{Ph} using the reciprocity relation (see Sect. 2.5). In Fig. 4.3, the calculated $V_{OC}^{Rad}(\epsilon_{min})$ *versus* the photon energy is shown. The truncated V_{OC}^{Rad} first decreases with reducing photon energy, reaching a plateau at energies near the CT state absorption, and then again decreases to lower values. The experimental V_{OC} values is shown as an horizontal dashed line in the plot, and the ΔV_{OC}^{NR} values determined at the CT state energy (the plateau) is indicated adjacent to the double headed arrows. This compares well with non-radiative losses determined from experimentally measured EQE$_{LED}$ which is provided as legend in the plot. However, if the low-energy sub-gap feature

Fig. 4.3 The calculated $V_{OC}^{Rad}(\epsilon_{min})$ (solid purple curve) as a function of the photon energy and the experimental V_{OC} measured at 1 sun illumination (dashed lines) are shown. The corresponding ΔV_{OC}^{NR}, calculated from the measured EQE_{LED} (using $q\Delta V_{OC}^{NR} = -k_B T \ln EQE_{LED}$), is shown as legend

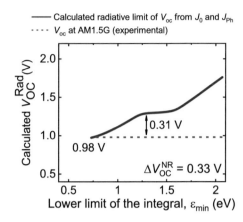

is included in the analysis (i.e. the truncation is reduced to the full measurement range of the ultra-sensitive EQE_{PV}), the non-radiative loss tend to zero in direct contradiction with reciprocity. In order to understand the origin of this contradiction, we need to first identify the origin of the low-energy sub-gap absorption features and the mechanism of charge generation through them.

In Figs. 4.4, 4.5, 4.6 and 4.7 similar measurements and analyses are shown for various devices including organic semiconductor solar cells, both fullerene and non-fullerene based. For comparison two exemplary inorganic devices, a c-Si solar cell and a germanium photodiode, are also included (see Fig. 4.7) [24, 25]. The EQE_{PV} spectra are sorted with respect to the V_{OC} of the devices, from highest (1.17 V) to lowest (0.21 V). The sub-gap absorption features are universally present in all organic semiconductor systems studied in this work result in the same contradiction (The details of device fabrication can be found in Appendix).

For the c-Si solar cell and the germanium photodiode the truncated V_{OC}^{Rad} rapidly saturates to a constant value for energies below the bandgap, suggesting that radiative voltage losses are insensitive to sub-bandgap features. The corresponding non-radiative voltage losses (0.19 V for c-Si) are in excellent agreement with literature values [26].

4.3   Origin of the Low-Energy Sub-Gap Absorption Features in the EQE_{PV}

In Fig. 4.8, the ultra-sensitive EQE_{PV} spectrum of a solar cell based upon the well-understood donor-acceptor system PCDTBT:PC$_{70}$BM is presented. Two distinct absorption features are readily apparent in the sub-gap region. The first feature at an energy of around 1.5 eV has been previously attributed to the CT state absorption which is often described in terms of Marcus theory: $EQE_{(PV,CT)}(\epsilon) = g(\epsilon, \epsilon_{CT}, \lambda_{CT}, f_{CT})T(\epsilon)$ [27]. Here, $T(\epsilon)$ is the cavity (solar cell) spectral through-

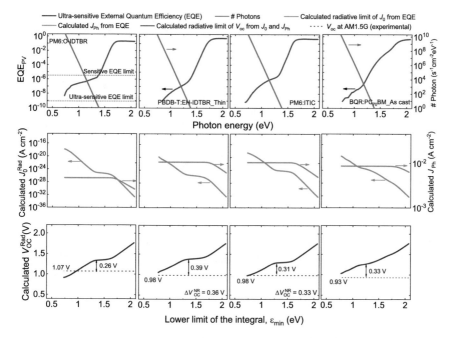

Fig. 4.4 In the upper row, the black curves represent the experimental EQE$_{PV}$ *versus* photon energy for organic solar cell devices comprising PM6:O-IDTBR, PBDB-T:EH-IDTBR, PM6:ITIC and as-cast BQR:PC$_{70}$BM. The corresponding ϕ_{BB} *versus* photon energy is plotted on the right axis (blue curve). In the middle row, The calculated $J_0^{Rad} = q \int_{\epsilon_{min}}^{\infty} EQE_{PV}(\epsilon) \times \phi_{BB}(\epsilon)d\epsilon$ (the green curve on the left axis) and $J_{Ph} = q \int_{\epsilon_{min}}^{\infty} EQE_{PV}(\epsilon) \times \phi_{Sun}(\epsilon)d\epsilon$ (the pink curve on the right axis) are shown *versus* the photon energy. In the lower row, the calculated $V_{OC}^{Rad}(\epsilon_{min})$ (solid purple curve) as a function of the photon energy and the experimental V_{OC} measured at 1 sun illumination (dashed lines) are shown. The corresponding ΔV_{OC}^{NR}, calculated from the measured EQE$_{LED}$ (using $q\Delta V_{OC}^{NR} = -k_B T \ln EQE_{LED}$), are shown as legends for PBDB-T:EH-IDTBR and PM6:ITIC devices

put while $g(\epsilon, \epsilon_{CT}, \lambda_{CT}, f_{CT})$ is given by

$$g(\epsilon, \epsilon_{CT}, \lambda_{CT}, f_{CT}) = \frac{f_{CT}}{\epsilon \sqrt{4\pi k_B T \lambda_{CT}}} \exp\left(-\frac{(\epsilon_{CT} + \lambda_{CT} - \epsilon)^2}{4\lambda_{CT} k_B T}\right). \qquad (4.1)$$

As discussed in Sect. 2.6, $g(\epsilon, \epsilon_{CT}, \lambda_{CT}, f_{CT})$ parametrises the CT state in terms of ϵ_{CT} which is the energy difference between the ground and excited state of the CT state, λ_{CT} which is the reorganization energy due to the formation of the CT state, and f_{CT} which is a measure of the strength of the donor-acceptor coupling and also proportional to the density of the CT states [8, 28, 29]. The CT state parameters ϵ_{CT}, λ_{CT} and f_{CT} can be approximated by fitting $g(\epsilon, \epsilon_{CT}, \lambda_{CT}, f_{CT})T(\epsilon)$ on the CT state region of the EQE$_{PV}$ (ϵ_{CT}, λ_{CT} and f_{CT} are free parameters of the fit) assuming

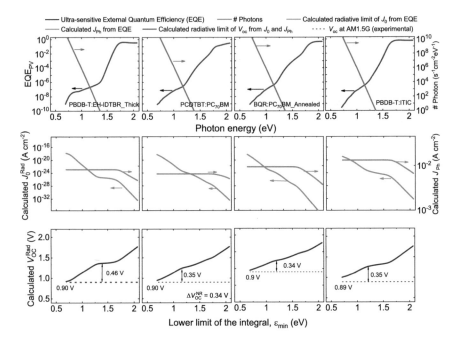

Fig. 4.5 In the upper row, the black curves represent the experimental EQE_{PV} *versus* photon energy for organic solar cell devices comprising PBDB-T:EH-IDTBR, PCDTBT:PC$_{70}$BM, annealed BQR:PC$_{70}$BM and, PBDB-T:ITIC. The corresponding ϕ_{BB} *versus* photon energy is plotted on the right axis (blue curve). In the middle row, The calculated $J_0^{Rad} = q \int_{\epsilon_{min}}^{\infty} EQE_{PV}(\epsilon) \times \phi_{BB}(\epsilon)d\epsilon$ (the green curve on the left axis) and $J_{Ph} = q \int_{\epsilon_{min}}^{\infty} EQE_{PV}(\epsilon) \times \phi_{Sun}(\epsilon)d\epsilon$ (the pink curve on the right axis) are shown *versus* the photon energy. In the lower row, the calculated $V_{OC}^{Rad}(\epsilon_{min})$ (solid purple curve) as a function of the photon energy and the experimental V_{OC} measured at 1 sun illumination (dashed lines) are shown. The corresponding ΔV_{OC}^{NR}, calculated from the measured EQE_{LED} (using $q\Delta V_{OC}^{NR} = -k_B T \ln EQE_{LED}$), is shown as legend for PCDTBT:PC$_{70}$BM device

$T(\epsilon)$ varies slowly with the photon energy for thin junctions. In the case of the PCDTBT:PC$_{70}$BM cell of Fig. 4.8, we find $\epsilon_{CT} = 1.48$ eV and $\lambda_{CT} = 0.35$ eV.

Apart from the CT states, a second absorption feature at low energies can be distinguished. Similar features have been previously observed by Street et al in PCDTBT: PC$_{70}$BM [25]. Here, we observe that the additional low-energy absorption features can also be accurately fitted with the Marcus formalism. This is to be expected considering that Marcus theory generally describes any type of charge transfer between weakly-coupled states undergoing non-adiabatic transitions. The corresponding energy and reorganization energy for this second low-energy sub-gap state, were found to be 0.74 eV and 0.56 eV, respectively. The energy of the low-energy (LE) sub-gap states appears to be exactly half of the CT state energy for PCDTBT:PC$_{70}$BM, suggesting they are associated with mid-gap states at the donor-acceptor interface.

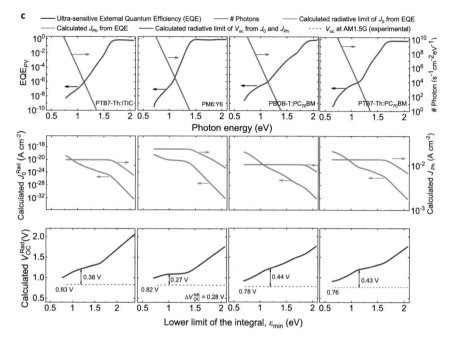

Fig. 4.6 In the upper row, the black curves represent the experimental EQE_PV *versus* photon energy for organic solar cell devices comprising PTB7-Th:ITIC, PM6:ITIC, PBDB-T:PC$_{70}$BM and PTB7-Th:PC$_{70}$BM. The corresponding ϕ_{BB} *versus* photon energy is plotted on the right axis (blue curve). In the middle row, The calculated $J_0^{Rad} = q \int_{\epsilon_{min}}^{\infty} EQE_{PV}(\epsilon) \times \phi_{BB}(\epsilon)d\epsilon$ (the green curve on the left axis) and $J_{Ph} = q \int_{\epsilon_{min}}^{\infty} EQE_{PV}(\epsilon) \times \phi_{Sun}(\epsilon)d\epsilon$ (the pink curve on the right axis) are shown *versus* the photon energy. In the lower row, the calculated $V_{OC}^{Rad}(\epsilon_{min})$ (solid purple curve) as a function of the photon energy and the experimental V_{OC} measured at 1 sun illumination (dashed lines) are shown. The corresponding ΔV_{OC}^{NR}, calculated from the measured EQE$_{LED}$ (using $q\Delta V_{OC}^{NR} = -k_B T \ln EQE_{LED}$), is shown as legend for PM6:Y6 device

By introducing a parameter "n" in the Marcus formula, replacing ϵ_{CT} with $\epsilon_t = \epsilon_{CT}/n$, f_{CT} with f_t and λ_{CT} with λ_t, we define $EQE_{PV,t}(\epsilon) = g(\epsilon, \epsilon_t, \lambda_t, f_t)$ to describe the low-energy sub-gap states in the EQE_{PV}, in which the energy of the low-energy sub-gap absorption relates to the CT state energy via the fitting parameter n. The total sub-gap region of the EQE_{PV} can then be described by

$$EQE_{PV} = EQE_{PV,CT} + EQE_{PV,t} = g(\epsilon, \epsilon_{CT}, \lambda_{CT}, f_{CT}) + g(\epsilon, \epsilon_t, \lambda_t, f_t). \quad (4.2)$$

This expression was then used to fit the entire sub-gap region of the EQE_{PV} for the organic semiconductor systems shown in Fig. 4.9 (fullerenes and non-fullerenes). The details of the fitting parameters are presented in Table 4.1. The values for n lie in the range of 1.6–2.1.

Table 4.1 Details of the Gaussian fits of the material systems presented in Fig. 4.6

Material system	Fitting parameters						
	ϵ_{CT} (eV)	λ_{CT} (eV)	f_{CT} (eV²)	ϵ_t (eV)	λ_t (eV)	f_t (eV²)	n
PM6:ITIC	1.61	0.15	9.0×10^{-2}	0.87	0.49	1.1×10^{-6}	1.85
PBDB-T:EH-IDTBR	1.62	0.17	5.9×10^{-3}	0.83	0.66	2.6×10^{-7}	1.93
BQR:PC$_{70}$BM	1.42	0.30	2.1×10^{-3}	0.75	0.60	3.1×10^{-7}	1.88
PBDB-T:ITIC	1.53	0.28	2.2×10^{-2}	0.87	0.45	2.9×10^{-7}	1.74
PTB7-Th:ITIC	1.44	0.47	9.0×10^{-2}	0.72	0.64	3.5×10^{-7}	1.99
PBDB-T:PC$_{70}$BM	1.39	0.56	2.5×10^{-2}	0.70	0.57	5.2×10^{-7}	1.96
PTB7-Th:PC$_{70}$BM	1.48	0.30	6.0×10^{-2}	0.82	0.32	3.1×10^{-7}	1.79
PCPDTBT:PC$_{70}$BM	1.33	0.29	1.6×10^{-2}	0.76	0.59	1.0×10^{-6}	1.74
PCDTBT:PC$_{70}$BM	1.48	0.35	4.9×10^{-3}	0.74	0.56	1.2×10^{-7}	2.02
PCDTBT:PC$_{70}$BM:0.1% m-MTDATA	1.47	0.34	5.5×10^{-3}	0.72	0.60	3.8×10^{-7}	2.04
PCDTBT:PC$_{70}$BM :1% m-MTDATA	1.47	0.34	3.6×10^{-3}	0.85	0.50	1.5×10^{-6}	1.72

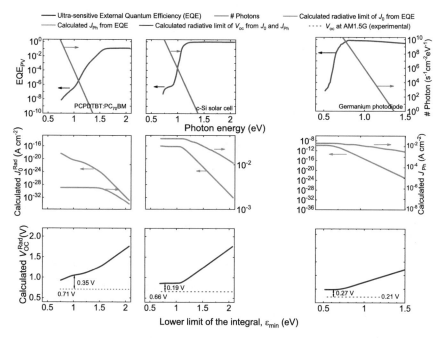

Fig. 4.7 In the upper row, the black curves represent the experimental EQE$_{PV}$ *versus* photon energy for an organic solar cell device comprising PCPDT-BT:PC$_{70}$BM, an inorganic c-Si solar cell and, a Germanium photodiode. The corresponding ϕ_{BB} *versus* photon energy is plotted on the right axis (blue curve). In the middle row, The calculated $J_0^{Rad} = q \int_{\epsilon_{min}}^{\infty} EQE_{PV}(\epsilon) \times \phi_{BB}(\epsilon)d\epsilon$ (the green curve on the left axis) and $J_{Ph} = q \int_{\epsilon_{min}}^{\infty} EQE_{PV}(\epsilon) \times \phi_{Sun}(\epsilon)d\epsilon$ (the pink curve on the right axis) are shown *versus* the photon energy. In the lower row, the calculated $V_{OC}^{Rad}(\epsilon_{min})$ (solid purple curve) as a function of the photon energy and the experimental V_{OC} measured at 1 sun illumination (dashed lines) are shown. For the c-Si solar cell and the Germanium photodiode the truncated V_{OC}^{Rad} rapidly saturates to a constant value for energies below the bandgap. The corresponding ΔV_{OC}^{NR} are in excellent agreement with literature values calculated using $q \Delta V_{OC}^{NR} = -k_B T \ln EQE_{LED}$ (the value of 0.19 V has been reported for c-Si)

We note, however, that the fittings are also sensitive to changes in the thickness of the different layers within the solar cell stack due to optical interference effects as shown by Kaiser et al. [27]. For example, by varying the thickness of the PCDTBT:PC$_{70}$BM active layer in the range of 56–113 nm, n varies in the range of 1.73–2.07 (see Fig. 4.10).

To further clarify whether these low-energy sub-gap states are associated with (bound) charges in mid-gap states, we intentionally increased the trap density by introducing a small amount of m-MTDATA into the active layer of PCDTBT: PC$_{70}$BM. m-MTDATA is a small molecule donor with a shallow HOMO level in the gap of PCDTBT:PC$_{70}$BM. The energetics are schematically represented in Fig. 4.11a. The normalized EQE$_{PV}$ of the devices with 0.1 and 1% m-MTDATA (by molar content of PCDTBT) together with the device with no additive are shown in Fig. 4.11b.

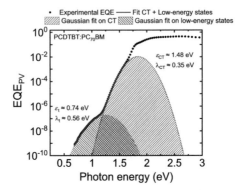

Fig. 4.8 The dotted black line represent the experimental EQE_{PV} of a PCDTBT: $PC_{70}BM$ solar cell *versus* photon energy. The two Gaussian fits, corresponding to CT state absorption and the low-energy sub-gap absorption (shaded area), together with fitting parameters are presented. The energy of the low-energy sub-gap state is half of the CT state energy

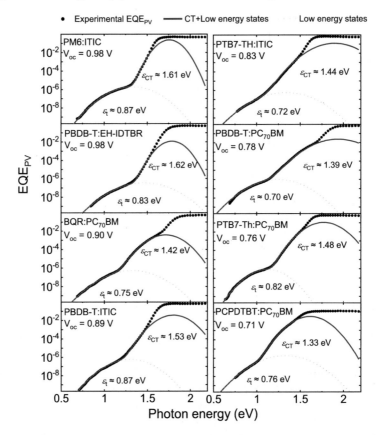

Fig. 4.9 The dotted black curves represent the experimental EQE_{PV} of several organic solar cells. The two Gaussian fits suggest that the energy of low-energy sub-gap state is almost half of the CT state energy for all of the presented devices

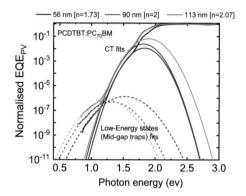

Fig. 4.10 The experimental EQE$_{PV}$ of PCDTBT:PC$_{70}$BM solar cells *versus* photon energy with different thicknesses of the active layer. The fitting parameter n is affected by the thickness of the active layer

It can be seen that by increasing the amount of m-MTDATA the EQE$_{PV}$ in the low-energy sub-gap region increases, with the normal CT state feature unchanged, as clearly apparent from the fits. The thickness of the active layer (and all the other layers) was kept constant in all devices, indicating that the increase of the low-energy sub-gap signal is caused solely by the increased trap density in the active layer. This supports the hypothesis that the low-energy sub-gap feature is associated with charges trapped in mid-gap states. It should be stressed that this does not necessarily exclude the presence of other trap states. However, in accordance with SRH statistics, it is expected that the contribution from states in the middle of the gap will be dominant [26, 30].

4.4   Charge Generation and Recombination via Mid-Gap States

Irrespective of the exact origin of the low-energy mid-gap states, it is clear that absorption into these states contributes to photocurrent in a similar manner to intermediate-gap solar cells, however, with negligible contribution to the total photocurrent [24, 25]. Charge generation through mid-gap states can be explained by a process known as optical release (or photoionization) [30, 31]. Figure 4.12 shows a schematic diagram of the energy levels at the donor-acceptor interface of an organic solar cell. The energy levels of the acceptor LUMO (lowest unoccupied molecular orbital) and the donor HOMO are denoted by $\epsilon_{LUMO,A}$ and $\epsilon_{HOMO,D}$, respectively. The energy level of the mid-gap state is assumed to be close to the middle of the gap. An electron in the HOMO level of the donor (CT ground state) absorbs a low-energy photon (lower than the energy needed for CT state excitation) and is promoted to a state in the middle of the gap, resulting in the formation of a mid-gap state. The excited

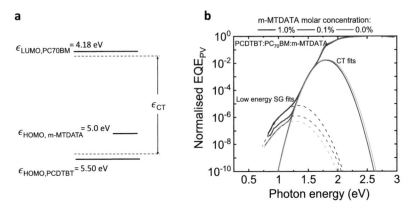

Fig. 4.11 **a** The schematic energy level diagram of PCDTBT:PC$_{70}$BM at the donor:acceptor interface and the HOMO level of m-MTDATA, functioning as a trap, is shown. **b** EQE$_{PV}$ of devices comprising PCDTBT:PC$_{70}$BM:(1% by mole) m-MTDATA and PCDTBT:PC$_{70}$BM: (0.1% by mole) m-MTDATA and PCDTBT:PC$_{70}$BM: (0% by mole) m-MTDATA is plotted *versus* photon energy. By adding more m-MTDATA (traps) the CT state parameters remain unchanged while the EQE$_{PV}$ signal in the low-energy sub-gap (SG) region increases

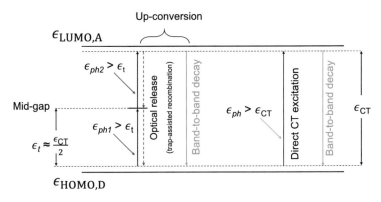

Fig. 4.12 Schematic energy level diagram at the donor-acceptor interface including mid-gap trap states. The associated generation and recombination routes are shown by upwards and downwards arrows, respectively

(trapped) electron in the mid-gap state can then be further released (from the trap) to the acceptor LUMO and thus contribute to charge generation if it absorbs a photon with energy higher than the trap energy depth ($\epsilon_{LUMO,A} - \epsilon_t$ offset). Note that this photon energy can be much lower than the CT state energy but needs to be large enough to promote the electron from the trap state into the acceptor LUMO.

In Fig. 4.12 the corresponding recombination pathways in the sub-gap region are shown schematically with downwards arrows. Note that the CT state decay (band-to-band recombination) can be both radiative and non-radiative [32]. According to the Franck Condon principle, the spectral position of the CT state photolumines-

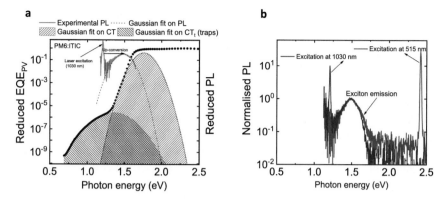

Fig. 4.13 a Reduced EQE$_{PV}$ of a PM6:ITIC device is plotted on the left axis (dotted curve). On the right axis the reduced PL of the same material system, excited at 1.2 eV (1030 nm), is plotted (dashed curve). The PL of the excited low-energy trap states emit at higher energies, consistent with optical release and subsequent photon up-conversion. **b** The normalised photoluminescence (PL) of the PM6:ITIC blend, excited at 2.4 eV (515 nm), is plotted *versus* photon energy (eV). The PL of excited excitons (blue line) is mainly emitted at lower energies (CT states), while the shoulder at 1.4 eV corresponds to the PL of the lowest lying excitons. The PL of excited low-energy trap states (purple line), on the other hand, emits only at 1.5 eV (CT state energy) which is consistent with optical release and subsequent photon up-conversion

cence (PL) (or equivalently EL in a full device) will be red-shifted relative to the absorption and the peak position can be described by $\epsilon^{PL}_{Peak,CT} = \epsilon^{Abs}_{Peak,CT} - 2\lambda_{CT}$ (in accordance with Marcus theory). Similar to CT states, which present a band-to-band recombination channel, mid-gap states can act as recombination centres presenting a trap-assisted recombination channel. We emphasize that, in this picture, each transition (from ground state to trap state and from trap state to CT state) may decay either radiatively or non-radiatively. In order to confirm whether the optical release mechanism via mid-gap states is operational, and according to the rationale above, we next investigated the recombination processes associated with these transitions. For this we utilised the PM6:ITIC system which has measurable and clearly identifiable PL (see Appendix for the details of the PL measurements). On the left axis of Fig. 4.13a the reduced EQE$_{PV}$ (i.e. EQE$_{PV}$ times the energy ϵ) of a PM6:ITIC device is shown, along with the corresponding Gaussian fits. On the left axis of the same plot the reduced PL spectrum (PL divided by ϵ) of a thin film of PM6:ITIC on glass, excited at 1.2 eV (1030 nm), is presented. As the CT state energy for this blend is about 1.6 eV which means that CT states absorb at a wavelength of about 775 nm, the pump beam at 1030 nm will exclusively excite the mid-gap states. However, the PL peak from this excitation appears at 1.47 eV which corresponds to energies where we observe the peak of the CT state PL when pumped at 515 nm (see Fig. 4.13b). This observation can only be explained in terms of a photon up-conversion process in which the sequential absorption of two low-energy photons ultimately generates a free electron-hole pair (CT state) which, upon recombining, emits a photon with higher energy.

4.4.1 Modified Shockley Read Hall (SRH) Theory

The original SRH model (Sect. 2.4.3) only accounts for the non-radiative transition of electrons (considering electron trap state) from the conduction band to trap states and from trap state to valence band. However, since the mid-gap states are detected in EQE_{PV} spectrum they are absorptive and in accordance with detail balance an absorptive state must also be radiative. The generation and recombination rates involving optical generation and radiative transitions of free electrons and holes taking place via trap states can be understood in terms of modified SRH statistics [30]. After accounting for radiative transitions, the modified SRH net generation-recombination rate via traps reads

$$U_{SRH}^{Mod} = \frac{\tilde{c}_n \tilde{c}_p N_t [np - n_1^{**} p_1^{**}]}{\tilde{c}_n [n + n_1^{**}] + \tilde{c}_p [p + p_1^{**}]}, \tag{4.3}$$

where n and p is the free electron and hole density, respectively, N_t is the trap density, while $\tilde{c}_{n(p)} = c_{n(p)} + r_{n(p)}$ is the coefficient for the transition of an electron (hole) between trap state and the conduction (valence) level being composed of radiative and non-radiative components $r_{n(p)}$ and $c_{n(p)}$, respectively. Here,

$$n_1^{**} = \frac{c_n}{c_n + r_n} \left(n_1 + \frac{G_n^{opt}}{c_n N_t} \right), \tag{4.4}$$

$$p_1^{**} = \frac{c_p}{c_p + r_p} \left(p_1 + \frac{G_p^{opt}}{c_p N_t} \right), \tag{4.5}$$

with $n_1 = N_C \exp\left(\frac{\epsilon_t - \epsilon_c}{k_B T}\right)$ and $p_1 = N_V \exp\left(\frac{\epsilon_v - \epsilon_t}{k_B T}\right)$, while G_n^{opt} and G_p^{opt} are the maximum optical generation rates for electrons and holes via traps, respectively, both depending linearly on the light intensity; ϵ_t is the energy of the trap state, ϵ_c is the energy of the conduction level, and ϵ_v is the energy of the valence level. Note that the conduction and valence level correspond to acceptor LUMO and donor HOMO levels, respectively. In accordance with detailed balance, we furthermore have [30]

$$r_n = \frac{1}{n_1} \int_0^\infty \sigma_n^{opt}(\epsilon) \phi_{BB}(\epsilon) d\epsilon, \tag{4.6}$$

$$r_p = \frac{1}{p_1} \int_0^\infty \sigma_p^{opt}(\epsilon) \phi_{BB}(\epsilon) d\epsilon, \tag{4.7}$$

where $\sigma_{p(n)}^{opt}$ is the corresponding absorption cross section for electrons (holes) and $\phi_{BB}(\epsilon)$ is the black-body spectrum of the environment. Finally, the associated net recombination-generation current density via traps is given by

$$J_{SRH} = q \int_0^d U_{SRH}^{Mod} dx, \tag{4.8}$$

where d is the active layer thickness.

4.4.1.1   Derivation of the Dark Current Density

For transitions predominately taking place via mid-gap states, we expect $n_1 = p_1 = n_i$ and $n \approx p \approx n_i \exp(qV/2K_BT)$, where n_i is the intrinsic carrier density. Then, assuming $\tilde{c}_n = \tilde{c}_p = \tilde{c}$ and that non-radiative transitions dominate over raditive ones (i.e. $c_{n(p)} \gg r_{n(p)}$), the associated current density in the dark ($G_n^{opt} = G_p^{opt} = 0$) simplifies as

$$J_{SRH} = J_2 = J_{02} \left[\exp\left(\frac{qV}{2k_BT}\right) - 1 \right] \tag{4.9}$$

with $J_{02} = q\tilde{c}N_t n_i d/2$ being the corresponding dark saturation current density. Furthermore, we assume optical transitions via mid-gap states to be governed by Marcus-type charge transfer with $\sigma_n^{opt} = \sigma_p^{opt} = \sigma_t^{opt}$ Accordingly, an absorption cross section of the form

$$\sigma_t^{opt}(\epsilon) = \frac{f_{\sigma t}}{\epsilon\sqrt{4\pi k_BT\lambda_t}} \exp\left(-\frac{(\epsilon_t + \lambda_t - \epsilon)^2}{4\lambda_t k_BT}\right), \tag{4.10}$$

is expected. Here, $f_{\sigma t}$ is a prefactor that depends on the oscillator strength. On the other hand, the absorption coefficient for optical trap generation can be expressed as $\alpha_t = \tilde{f}_t \sigma_t N_t$, where $\tilde{f}_t$ is the occupancy of the trap states which for mid-gap states is $\tilde{f}_t = 1/2$. Then, after noting that for weakly absorbing states the EQE_{PV} may be approximated as $EQE_{PV,t} = \alpha_t d$, we finally obtain

$$J_{02} = \frac{q}{EQE_{LED,t}} \int_0^\infty EQE_{PV,t}(\epsilon)\phi_{BB}(\epsilon)d\epsilon, \tag{4.11}$$

where $EQE_{LED,t} = \frac{r_{n(p)}}{c_{n(p)} + r_{n(p)}}$ denotes the radiative efficiency of the states, while

$$EQE_{PV,t} = \frac{f_t}{\epsilon\sqrt{4\pi k_BT\lambda_t}} \exp\left(-\frac{(\epsilon_t + \lambda_t - \epsilon)^2}{4\lambda_t k_BT}\right) \tag{4.12}$$

with $f_t = f_{\sigma t}N_t d/2$.

It should be noted that J_2 is generally dependent on the light intensity (via G_n^{opt} and G_p^{opt}). However, owing to the extremely weak absorption of traps in our case, the rate U_{SRH}^{Mod} is dominated by injected carriers in forward bias ($n \gg n_1^{**}$); hence, the expressions for J_2 and J_{02} derived for dark conditions remain valid under open-circuit conditions (at 1 sun).

Fig. 4.14 The photocurrent *versus* intensity measurement at excitation wavelength of 1550 nm for a PCDTBT:PC$_{70}$BM device at short-circuit. The photocurrent is in this case exclusively induced by optical generation via mid-gap states, showing a linear intensity dependence, as expected from modified SRH theory

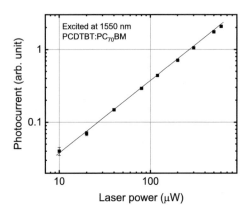

4.4.1.2   Conditions When Optical Generation via Mid-gap States Dominates: EQE$_{PV}$ *versus* PL

The optical generation via traps becomes dominant under special conditions when the influence of injected carriers from the contacts is negligible and the generation of free charge carriers by direct optical transitions are absent. For mid-gap states, assuming thermal generation to be negligible ($G_t^{opt}\tau \gg n_i$), the following simplified rate equation for free charge carriers can be obtained

$$\frac{n}{t_{col}} = \frac{(G_t^{opt}\tau)^2 - n^2}{2\tau(n + G_t^{opt}\tau)} - \beta n^2 \tag{4.13}$$

assuming $n = p$, $G_n^{opt} = G_p^{opt} = G_t^{opt}$ and $\tau = (\tilde{c}N_t)^{-1}$, where $\tilde{c} = \tilde{c}_n = \tilde{c}_p \gg r_n = r_p$; moreover, t_{col} is the charge collection time and β is the band to band recombination coefficient. Here, the term on the left-hand-side represents the charge extraction rate, while the first and second term on the right-hand-side corresponds to trap-assisted net generation-recombination rate (based on modified SRH theory) and the band to band recombination rate, respectively.

Under short-circuit conditions, the carrier density is expected to be small and the recombination terms negligible. Subsequently, the short-circuit current density, $J_{sc} \propto n/t_{col}$, takes the form

$$J_{sc} \propto G_t^{opt} \tag{4.14}$$

being linear with the light intensity. Hence, we expect the photocurrent induced by mid-gap states to be linear with light intensity at short-circuit (at low generation levels). This is also seen experimentally in Fig. 4.14.

In PL measurements, on the other hand, charge-extracting electrodes are absent, corresponding to $t_{col} = \infty$. Under these conditions, everything that is generated ultimately recombines; after neglecting third-order terms for the carrier density, we find

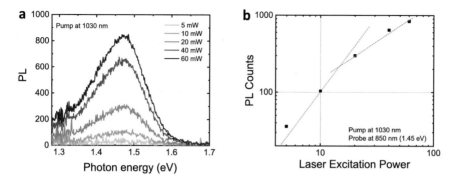

Fig. 4.15 **a** PL spectra measured with different pump intensities plotted versus photon energy. **b** PL spectra count at the peak (1.45 eV) is plotted *versus* laser excitation power. The increase of the laser power leads to a quadratic growth of the PL intensity at lower power, while a linear dependence is observed at higher power. This behaviour is consistent with the behaviour expected from modified SRH theory (see above), strongly supporting the presence of optical release

$$n^2 \approx \frac{(G_t^{\mathrm{opt}}\tau)^2}{2\beta G_t^{\mathrm{opt}}\tau^2 + 1} \tag{4.15}$$

subsequently, for PL originating from band to band recombination (PL $\propto \beta n^2$), we expect a quadratic intensity dependence [PL $\propto (G_t^{\mathrm{opt}})^2$] at low intensities and a linear intensity dependence [PL $\propto G_t^{\mathrm{opt}}$] at high intensities. This explains the experimentally observed behavior in Fig. 4.15. The observed up-conversion is, therefore, an indication of the optical release mechanism.

4.5 The Two-Diode Model and the Origin of the Ideality Factor in OSCs

A direct consequence of the presence of the partially radiative trap states is that the reciprocity relation no longer applies in the form shown which is based upon a linear extrapolation from equilibrium to quasi-equilibrium [32]. Instead, the generation-recombination channel via traps needs to be described separately [33]. The total bulk recombination current, taking place via CT and the low-energy sub-gap channels, is described by two parallel currents J_1 and J_2 being the CT-induced recombination current and the trap-induced recombination current, respectively. The total dark current is given by

$$J_D^{\mathrm{tot}} = J_1 + J_2 + J_{\mathrm{Shunt}}. \tag{4.16}$$

This system can be described in terms of the equivalent circuit Fig. 4.16.

Accordingly, the diode current J_1 only involves the direct recombination between free electrons and holes, being governed by their respective quasi-Fermi levels (ϵ_{Fn}

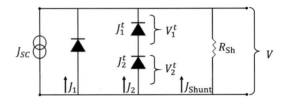

Fig. 4.16 The equivalent circuit of the two-diode model. The diode current is given by the contributions of two parallel recombination currents J_1 (from CT state recombination) and J_2 (from trap-assisted recombination); J_{Shunt} is the leakage current caused by an external shunt resistance R_{Sh}

for electrons, and ϵ_{Fp} for holes). The associated current-voltage (J-V) characteristics is governed by the quasi-Fermi level difference $qV = \epsilon_{Fn} - \epsilon_{Fp}$ between electrons and holes (at the electrodes); hence,

$$J_1 = J_{01}\left[\exp\left(\frac{qV}{k_B T}\right) - 1\right] \tag{4.17}$$

where, in accordance with the reciprocity principle, J_{01} is given by

$$J_{01} = \frac{q}{EQE_{LED,CT}} \int_0^\infty EQE_{PV,CT}(\epsilon)\phi_{BB}(\epsilon)d\epsilon. \tag{4.18}$$

On the other hand, the recombination (and dark generation) of free electrons and holes via trap states is composed of a two-step process: (**i**) the transition involving a free hole and a trap, and (**ii**) the transition involving a free electron and a trap. (see Fig. 4.17) Furthermore, in accordance with SRH statistics [26], trapped carriers occupying the mid-gap states can be described by their own quasi-fermi level ϵ_{Ft}. Subsequently, the diode current J_2 induced by trap-assisted recombination between free electrons and holes can be described by two diode components which are in series with each other: the first diode current J_1^t being governed by the quasi-Fermi level difference $qV_1^t = \epsilon_{Ft} - \epsilon_{Fp}$ (i), and the second J_2^t by $qV_2^t = \epsilon_{Fn} - \epsilon_{Ft}$ (ii); hence,

$$J_1 = J_{01}^t\left[\exp\left(\frac{qV_1^t}{k_B T}\right) - 1\right] \tag{4.19}$$

and

$$J_2 = J_{02}^t\left[\exp\left(\frac{qV_2^t}{k_B T}\right) - 1\right], \tag{4.20}$$

where J_{01}^t and J_{02}^t are the corresponding dark saturation currents associated with processes (i) and (ii), respectively. For the case when mid-gap traps are dominant, we expect $V_1^t = V_2^t = V/2$. Then, noting that $J_2 = J_1^t = J_2^t$ (because of the series connection), it follows that

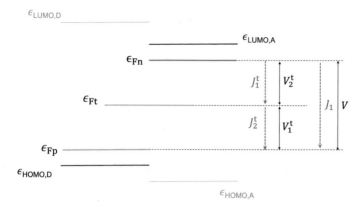

Fig. 4.17 The schematic energy levels at the donor-acceptor interface are shown together with the quasi-Fermi levels of electrons (ϵ_{Fn}), holes (ϵ_{Fp}) and the traps (ϵ_{Ft}). The band to band recombination current J_1 and the trap assisted recombination currents J_1^t and J_2^t are shown with downwards arrows

$$J_2 = J_{02}\left[\exp\left(\frac{qV}{2k_BT}\right) - 1\right] \tag{4.21}$$

with J_{02} being the corresponding dark saturation current for this recombination channel, which in general is composed both radiative and non-radiative transitions. An explicit expression for J_{02} can be derived based on the modified SRH theory which takes radiative transitions via traps into account Eq. 4.11.

As the result, the total dark current (Eq. 4.16) can be presented as:

$$J_D^{tot} = J_1 + J_2 + J_{Shunt} = J_{01}\left[\exp\left(\frac{qV}{k_BT}\right) - 1\right] + J_{02}\left[\exp\left(\frac{qV}{2k_BT}\right) - 1\right] + J_{Shunt} \tag{4.22}$$

where $J_{01} = (J_{01}^{Rad})/EQE_{LED,CT}$ and $J_{02} = (J_{02}^{Rad})/EQE_{LED,t}$ are the corresponding dark saturation currents of diode 1 and diode 2, respectively, with the corresponding radiative contributions $J_{01}^{Rad} = q\int_0^\infty EQE_{PV,CT}(\epsilon)\phi_{BB}(\epsilon)d\epsilon$ and $J_{02}^{Rad} = \int_0^\infty EQE_{PV,t}(\epsilon)\phi_{BB}(\epsilon)d\epsilon$. Finally, $EQE_{LED,CT}$ and $EQE_{LED,t}$ are the respective quantum efficiencies for the electroluminescence of CT states and mid-gap states, describing their radiative efficiencies.

In accordance with Eq. 4.22, the total diode current is thus given by a combination of two diode components, one with an ideality factor $n_{id} = 1$ and the other with an ideality factor $n_{id} = 2$. We note, however, that the current eventually becomes transport-limited (and/or series-resistance-limited) at larger voltages when the built-in voltage of the cell is approached. Figure 4.18 show the experimental dark J-V of the BQR:PC$_{70}$BM and PM6:Y6 systems, respectively (see Appendix for the details of the dark J-V measurements). The corresponding values for J_{01}^{Rad} and J_{02}^{Rad} are shown in the insets and were calculated based on the EQE_{PV}.

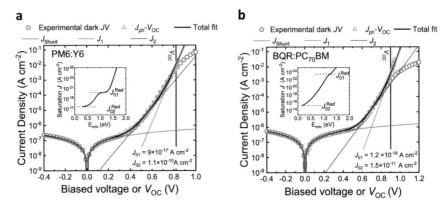

Fig. 4.18 **a** PM6:Y6 and **b** BQR:PC$_{70}$BM are shown in semi-log plots (empty circled curve). The inset plots show the integrated J_{01}^{Rad} and J_{02}^{Rad}, being the radiative dark saturation currents of CT states and mid-gap traps, respectively, as calculated from EQE$_{PV}$. The values of J_{01}^{Rad} and J_{02}^{Rad}, respectively, corresponding to the dashed line in panels a and b are 5.3×10^{-21} A cm^{-2} and 4.5×10^{-18} A cm^{-2} for PM6:Y6 and 9.8×10^{-24} A cm^{-2} and 1.4×10^{-15} A cm^{-2} for BQR:PC$_{70}$BM. Equation 4.22 is used to fit the J-V curves and the contributions of J_{Shunt}, J_1, J_2 are shown in the plots (purple, red and blue curves, respectively). For both cases, the diode 1 governs the total dark current $V = V_{OC}$ if incident light irradiance is large enough. See the vertical line, which represents V_{OC} measured at 1 sun

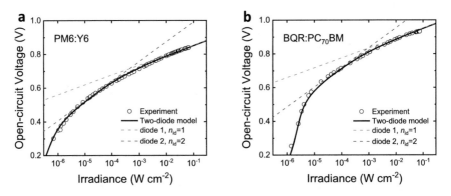

Fig. 4.19 $J_{Ph} - V_{OC}$ curves in orange, for **a** PM6:Y6 and **b** BQR:PC$_{70}$BM as obtained from intensity dependent V_{OC} measurements are provided to compare with the dark J-V

For comparison, we have also included experimental $J_{Ph} - V_{OC}$ curves, corresponding to ideal J-V curves free of series resistance and transport limitations [34]. $J_{Ph} - V_{OC}$ curves were obtained from corresponding intensity dependent V_{OC} measurements, as shown in Fig. 4.19. Note that the photocurrent J_{Ph} is directly proportional to the light intensity (see Appendix for details of $J_{Ph} - V_{OC}$ measurement).

Subsequently, Eq. 4.22 was used to fit on the J-V curves using EQE$_{LED,CT}$, EQE$_{LED,t}$ and R_{Sh} as fitting parameters (see Table 4.2). As a result, the total dark J-V curve can be described by three distinct current components J_1, J_2 and J_{Shunt}.

Table 4.2 Details of the two diode model fits presented in Fig. 4.18

		Fitting parameters		
		R_{Shunt} Ωcm^2	$EQE_{LED,CT}$	$EQE_{LED,t}$
Material system	PM6:Y6	1.8×10^6	5.6×10^{-5}	4.1×10^{-8}
	BQR:PC$_{70}$BM	7.4×10^5	8.2×10^{-6}	9.3×10^{-5}

At open circuit under 1 sun illumination, the total current is mainly dominated by J_1 which implies that the radiative limit of the V_{OC} is determined by J_{01}. However, the complete J-V curve of the cells cannot be explained by J_{01} alone. This resolves the apparent contradiction of our initial observations regarding the detailed balance. Furthermore, the extracted $EQE_{LED,CT}$ are in good agreement with those expected from Fig. 4.1. We note that it is nearly impossible to directly measure $EQE_{LED,t}$.

Our findings also provide compelling evidence for the origin of the ideality factor in organic photovoltaic devices when bulk recombination is dominant. Ideality factors ranging between 1 and 2 have been frequently observed in organic solar cells, however, the underlying mechanism has remained under debate [34, 35]. In light of the two-diode model the ideality factor is determined by the competition between CT state recombination, with $n_{id} = 1$, and trap-assisted recombination via mid-gap states with $n_{id} = 2$. This is further demonstrated in Fig. 4.13, showing excellent agreement between the experimental V_{OC} results and the two-diode model. Note in particular the gradual transition, taking place over several orders of magnitudes in intensity, from $n_{id} = 2$ to $n_{id} = 1$ as the intensity is increased. This slow transition ultimately manifests itself as an apparent arbitrary non-integer ideality factor >1 in the experiments with limited dynamic range fitted with a one-diode equation. Our data here show that the ideality factor is not a constant and undergoes a transition from 1 to 2 as the V_{OC} changes. Note that the V_{OC} is limited by shunt effects at low intensities.

It should be stressed that the CT recombination current J_1 is composed of a radiative and a non-radiative component both described by an ideality factor of one. This is in accordance with recent findings suggesting that non-radiative recombination via CT states predominately limits the V_{OC} of organic solar cells at 1 sun [32, 36]. Since radiative and non-radiative recombination via CT states are both initiated by the encounter of the same type of separate charge carriers, they are also expected to have the same ideality factor ($n_{id} = 1$). In other words, the CT contribution in the EQE_{PV} reflects states that recombine with $n_{id} = 1$, whereas the mid-gap state contribution reflects states that recombine with $n_{id} = 2$. This trade-off is evident from voltage-dependent EQE_{LED} shown in Fig. 4.20 (see Appendix for the details of the EQE_{LED} measurements). We note that a non-integer ideality factor $n_{id} > 1$ can also arise from trap-assisted recombination via exponential tail states [35]. If this type of non-radiative recombination channel is present, then a corresponding radiative component with $n_{id} > 1$, reflected by a corresponding exponential tail in the EQE_{PV}, is to be expected as well. This has been previously observed in inorganic solar cells such as a-Si [33, 37]. For the organic systems studied in this work, however, no such

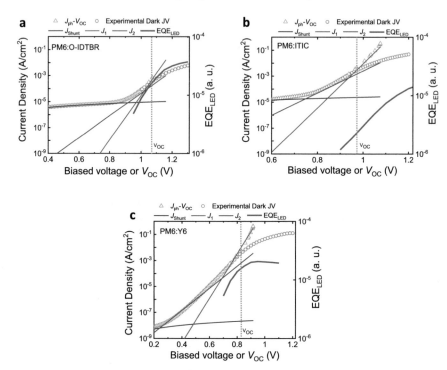

Fig. 4.20 Two-diode fittings are performed on three exemplary organic solar cells **a** PM6:O-IDTBR **b** PM6:ITIC and **C** PM6:Y6. The V_{OC} is marked with a vertical dashed line on the plots. In all three systems the EQE_{LED} (green curves) at lower voltages is clearly voltage dependent due to the trade-off between radiative J_1 (red curve) and non-radiative J_2 (blue curve). In PM6:Y6 the variations in the EQE_{LED} near V_{OC} is marginal where J_1 dominates the current. In PM6:O-IDTBR and PM6:ITIC J_1 and J_2 are approximately equally contributing to the dark current with the EQE_{LED} being significantly voltage dependent

tails can be distinguished from the ultra-sensitive EQE_{PV} spectra, suggesting that recombination through exponential tail states, if present, is negligibly small compared to the other recombination channels in these systems. Finally, we note that while mid-gap states do not appear to significantly affect the V_{OC} (at 1 sun) for the organic solar cells studied in this work, this may not always be the case depending on the cross-over voltage between J_1 and J_2 especially for thick junctions—a matter of significant importance for the viable scaling of organic solar cells.

4.6 Impact on the Detectivity of Organic Photodetectors

The origin of the dark current has important implications for photodiodes. The specific detectivity of a photodiode is define as

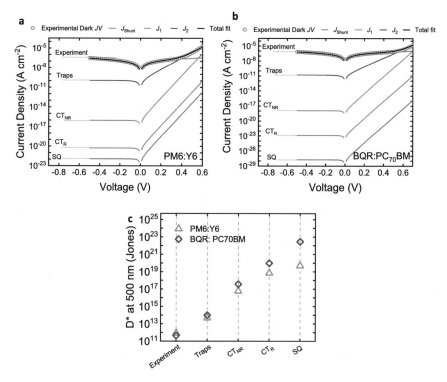

Fig. 4.21 The experimental dark J-V curves of **a** PM6:Y6 and **b** BQR:PC$_{70}$BM, including the reverse-bias region (photodetector mode), are shown. The dark J-V contributions from different limiting recombination processes: Shockley–Queisser (SQ), radiative limit of the CT states (CT$_R$), non-radiative CT states (CT$_{NR}$), and trap states are shown, respectively, from the bottom to the top of the plot. **c** The corresponding shot noise-limited specific detectivity at 500 nm calculated for the different dark saturation currents from panel **a** for PM6:Y6 (plotted in pink) and from panel **b** for BQR:PC$_{70}$BM (plotted in blue); the theoretical limit of the detectivity is determined by the recombination current via trap states

$$D^* = \frac{\lambda \sqrt{A}.\mathrm{EQE_{PV}}}{hc.i_{\mathrm{noise}}}, \tag{4.23}$$

where i_{noise} is commonly express as the summation of the shot noise (first term) and the thermal noise (second term) as follows [38, 39]

$$\langle i^2_{\mathrm{noise}} \rangle = 2qi_{dark} + \frac{4k_B T}{R_{\mathrm{Shunt}}}. \tag{4.24}$$

In a device with large R_{Shunt}, the dark saturation current defines the i_{noise} and consequently the specific detectivity for which information on a theoretical limit is still lacking in the case of organic photodetectors.

Figure 4.21 demonstrate the experimental dark J-V (circle) along with the calculated contributions from CT states J_{01} (red curve) and mid-gap states J_{02} (dark blue curve) to the total recombination current for two material systems PM6:Y6 and BQR:PC$_{70}$BM. For comparison, contributions from the corresponding radiative limit of the CT states J_{01}^{Rad} (green curve) and the Shockley–Queisser (SQ) limit (light blue curve), which only account for radiative exciton recombination (without considering the sub-gap region), have been included. Consequently, in the dark, the recombination via trap states is always dominant at low forward bias voltages and reverse bias. Note that the corresponding dark saturation current contribution for trap states is 10 orders of magnitude above the radiative CT limit and nearly 6 orders of magnitude above the non-radiative CT limit. This presents severe limitations on both the shot noise and the detectivity in organic photodiodes. In Fig. 4.21 the shot-noise limited specific detectivity (D^*) of PM6:Y6 and BQR:PC$_{70}$BM devices, calculated at a wavelength of 500 nm are shown for the different dark saturation current contributions (from panel a and b), namely: SQ limit [or the so-called background limited infrared photo-limit (BLIP)]; radiative CT state limit; non-radiative CT state limit; and trap state limit. These results demonstrate that mid-gap states set the thermodynamic limit of the detectivity in organic photodiodes which often operate in reverse bias where J_{02} dominates the dark saturation current. Critically, the resulting thermodynamic limit of D^* is several orders of magnitude lower than previous predictions neglecting the mid-gap states. Going forward, this may have a profound influence on our expectations of organic semiconductor photodetectors.

4.7   Conclusion

In conclusion, by utilizing ultra-sensitive photovoltaic external quantum efficiency measurements the presence of (partially bright) sub-gap states in organic semiconductor photovoltaic devices have been revealed. By considering these states in the V_{OC} calculation the conventional reciprocity relation between EQE$_{PV}$ and EQE$_{LED}$ fall into conflict as the predicted radiative limits of V_{OC} based upon reciprocity become non-physical. Furthermore, strong evidence are provided that these additional sub-gap features are associated with mid-gap states. Based on our findings, it has been shown that the dark J-V of organic photovoltaic devices can only be described with a two-diode model, providing an extension of the reciprocity principle, and reconciling detailed balance. Accordingly, two parallel recombination currents, one directly associated with CT states and the other with the trap states, determines the total bulk recombination current and hence the ideality factor in these systems. These currents also ultimately define the thermodynamic limit of V_{OC} in organic solar cells and the specific detectivity of organic photodiodes which is now found to be several of orders of magnitude lower than previously predicted.

References

1. Zarrabi, N., Sandberg, O., Zeiske, S., Li, W., Riley, D., Meredith, P., & Armin, A. (2020). Charge-generating mid-gap trap states define the thermodynamic limit of organic photovoltaic devices. *Nature Communications, 11*, 5567.
2. Shockley, W., & Queisser, H. (1961). Detailed balance limit of efficiency of p-n junction solar cells. *Journal of Applied Physics, 32*, 510–519.
3. Rau, U. (2007). Reciprocity relation between photovoltaic quantum efficiency and electroluminescent emission of solar cells. *Physical Review B, 76*, 085303.
4. Kirchartz, T., Taretto, K., & Rau, U. (2009). Efficiency limits of organic bulk heterojunction solar cells. *The Journal of Physical Chemistry C, 113*, 17958–17966.
5. Qi, B., & Wang, J. (2012). Open-circuit voltage in organic solar cells. *Journal of Materials Chemistry, 22*, 24315.
6. Tress, W., Marinova, N., Inganäs, O., Nazeeruddin, M., Zakeeruddin, S., & Graetzel, M. (2015). Predicting the open-circuit voltage of CH 3 NH 3 PbI 3 perovskite solar cells using electroluminescence and photovoltaic quantum efficiency spectra: The role of radiative and non-radiative recombination. *Advanced Energy Materials, 5*, 1400812. http://doi.wiley.com/10.1002/aenm.201400812
7. Fei, Z., Eisner, F., Jiao, X., Azzouzi, M., Röhr, J., Han, Y., Shahid, M., Chesman, A., Easton, C., McNeill, C., Anthopoulos, T., Nelson, J. & Heeney, M. An Alkylated Indacenodithieno[3,2- b]thiophene-based nonfullerene acceptor with high crystallinity exhibiting single junction solar cell efficiencies greater than 13.
8. Vandewal, K., Tvingstedt, K., Gadisa, A., Inganäs, O., & Manca, J. (2010). Relating the open-circuit voltage to interface molecular properties of donor: Acceptor bulk heterojunction solar cells. *Physical Review B, 81*, 125204.
9. Vandewal, K., Tvingstedt, K., Gadisa, A., Inganäs, O., & Manca, J. (2009). On the origin of the open-circuit voltage of polymer-fullerene solar cells. *Nature Materials, 8*, 904–909.
10. Vandewal, K., Gadisa, A., Oosterbaan, W., Bertho, S., Banishoeib, F., Van Severen, I., Lutsen, L., Cleij, T., Vanderzande, D., & Manca, J. (2008). The relation between open-circuit voltage and the onset of photocurrent generation by charge-transfer absorption in polymer: Fullerene bulk heterojunction solar cells. *Advanced Functional Materials, 18*, 2064–2070.
11. Liu, Q., Jiang, Y., Jin, K., Qin, J., Xu, J., Li, W., Xiong, J., Liu, J., Xiao, Z., Sun, K. & Others 18.
12. Zhan, L., Li, S., Lau, T., Cui, Y., Lu, X., Shi, M., Li, C., Li, H., Hou, J., & Chen, H. (2020). Over 17% efficiency ternary organic solar cells enabled by two non-fullerene acceptors working in an alloy-like model. *Energy & Environmental Science, 13*, 635–645.
13. Yuan, J., Zhang, Y., Zhou, L., Zhang, G., Yip, H., Lau, T., Lu, X., Zhu, C., Peng, H., Johnson, P., Leclerc, M., Cao, Y., Ulanski, J., Li, Y., & Zou, Y. (2019). Single-junction organic solar cell with over 15% efficiency using fused-ring acceptor with electron-deficient core. *Joule, 3*, 1140–1151.
14. Kirchartz, T., Mattheis, J., & Rau, U. (2008). Detailed balance theory of excitonic and bulk heterojunction solar cells. *Physical Review B, 78*, 235320.
15. Kirchartz, T., Nelson, J., & Rau, U. (2016). Reciprocity between charge injection and extraction and its influence on the interpretation of electroluminescence spectra in organic solar cells. *Physical Review Applied, 5*, 054003.
16. Melianas, A., Felekidis, N., Puttisong, Y., Meskers, S., Inganäs, O., Chen, W., & Kemerink, M. (2019). Nonequilibrium site distribution governs charge-transfer electroluminescence at disordered organic heterointerfaces. *Proceedings of The National Academy of Sciences, 116*, 23416–23425.
17. Zeiske, S., Kaiser, C., Meredith, P., & Armin, A. (2020). Sensitivity of sub-bandgap external quantum efficiency measurements of solar cells under electrical and light bias. *ACS Photonics, 7*, 256–264.

18. Vandewal, K., Benduhn, J., Schellhammer, K., Vangerven, T., Rückert, J., Piersimoni, F., Scholz, R., Zeika, O., Fan, Y., Barlow, S., Neher, D., Marder, S., Manca, J., Spoltore, D., Cuniberti, G., & Ortmann, F. (2017). Absorption tails of donor: C 60 blends provide insight into thermally activated charge-transfer processes and polaron relaxation. *Journal of The American Chemical Society, 139*, 1699–1704.
19. Nikolis, V., Benduhn, J., Holzmueller, F., Piersimoni, F., Lau, M., Zeika, O., Neher, D., Koerner, C., Spoltore, D., & Vandewal, K. (2017). Reducing voltage losses in cascade organic solar cells while maintaining high external quantum efficiencies. *Advanced Energy Materials, 7*, 1700855.
20. Chen, S., Wang, Y., Zhang, L., Zhao, J., Chen, Y., Zhu, D., Yao, H., Zhang, G., Ma, W., Friend, R., Chow, P., Gao, F., & Yan, H. (2018). Efficient nonfullerene organic solar cells with small driving forces for both hole and electron transfer. *Advanced Materials, 30*, 1804215.
21. Ziffer, M., Jo, S., Zhong, H., Ye, L., Liu, H., Lin, F., Zhang, J., Li, X., Ade, H., Jen, A., & Ginger, D. (2018). Long-lived, non-geminate, radiative recombination of photogenerated charges in a polymer/small-molecule acceptor photovoltaic blend. *Journal of The American Chemical Society, 140*, 9996–10008.
22. Ullbrich, S., Benduhn, J., Jia, X., Nikolis, V., Tvingstedt, K., Piersimoni, F., Roland, S., Liu, Y., Wu, J., Fischer, A., Neher, D., Reineke, S., Spoltore, D., & Vandewal, K. (2019). Emissive and charge-generating donor-acceptor interfaces for organic optoelectronics with low voltage losses. *Nature Materials, 18*, 459–464.
23. Kurpiers, J., Ferron, T., Roland, S., Jakoby, M., Thiede, T., Jaiser, F., Albrecht, S., Janietz, S., Collins, B., Howard, I., & Neher, D. (2018). Probing the pathways of free charge generation in organic bulk heterojunction solar cells. *Nature Communications, 9*, 2038.
24. Keevers, M., & Green, M. (1996). Extended infrared response of silicon solar cells and the impurity photovoltaic effect. *Solar Energy Materials and Solar Cells, 41–42*, 195–204.
25. Street, R., Krakaris, A., & Cowan, S. (2012). Recombination through different types of localized states in organic solar cells. *Advanced Functional Materials, 22*, 4608–4619.
26. Würfel, P., & Würfel, U. (2016). *Physics of solar cells: From basic principles to advanced concepts.* New York: Wiley.
27. Kaiser, C., Zeiske, S., Meredith, P., & Armin, A. (2020). Determining ultralow absorption coefficients of organic semiconductors from the sub-bandgap photovoltaic external quantum efficiency. *Advanced Optical Materials, 8*, 1901542.
28. Marcus, R. (1989). Relation between charge transfer absorption and fluorescence spectra and the inverted region. *The Journal of Physical Chemistry, 93*, 3078–3086.
29. Deibel, C., Strobel, T., & Dyakonov, V. (2010). Role of the charge transfer state in organic donor-acceptor solar cells. *Advanced Materials, 22*, 4097–4111.
30. Beaucarne, G., Brown, A., Keevers, M., Corkish, R., & Green, M. (2002). The impurity photovoltaic (IPV) effect in wide-bandgap semiconductors: An opportunity for very-high-efficiency solar cells? *Progress in Photovoltaics: Research and Applications, 10*, 345–353.
31. Hsieh, Y., & Card, H. (1989). Limitation to Shockley-Read-Hall model due to direct photoionization of the defect states. *Journal of Applied Physics, 65*, 2409–2415.
32. Benduhn, J., Tvingstedt, K., Piersimoni, F., Ullbrich, S., Fan, Y., Tropiano, M., McGarry, K., Zeika, O., Riede, M., Douglas, C., & Others,. (2017). Intrinsic non-radiative voltage losses in fullerene-based organic solar cells. *Nature Energy, 2*, 1–6.
33. Rau, U., Blank, B., Müller, T., & Kirchartz, T. (2017). Efficiency potential of photovoltaic materials and devices unveiled by detailed-balance analysis. *Physical Review Applied, 7*, 044016.
34. Tvingstedt, K., & Deibel, C. (2016). Temperature dependence of ideality factors in organic solar cells and the relation to radiative efficiency. *Advanced Energy Materials, 6*, 1502230.
35. Kirchartz, T., Deledalle, F., Tuladhar, P., Durrant, J., & Nelson, J. (2013). On the differences between dark and light ideality factor in polymer: Fullerene solar cells. *The Journal of Physical Chemistry Letters, 4*, 2371–2376.
36. Azzouzi, M., Yan, J., Kirchartz, T., Liu, K., Wang, J., Wu, H., & Nelson, J. (2018). Nonradiative energy losses in Bulk-Heterojunction organic photovoltaics. *Physical Review X, 8*, 031055.
37. Müller, T., Pieters, B., Kirchartz, T., Carius, R., & Rau, U. (2014). Effect of localized states on the reciprocity between quantum efficiency and electroluminescence in Cu(In, Ga)Se2 and Si thin-film solar cells. *Solar Energy Materials and Solar Cells, 129*, 95–103.

38. Armin, A., Hambsch, M., Kim, I., Burn, P., Meredith, P., & Namdas, E. (2014). Thick junction broadband organic photodiodes. *Laser & Photonics Reviews, 8*, 924–932.
39. Arquer, F., Armin, A., Meredith, P., & Sargent, E. (2017). Solution-processed semiconductors for next-generation photodetectors. *Nature Reviews Materials, 2*, 1–17.

Chapter 5
Relating Charge Transfer State Kinetics and Strongly Reduced Bimolecular Recombination in Organic Solar Cells

Abstract Significantly reduced bimolecular recombination relative to the Langevin recombination rate has been observed in a limited number of donor-acceptor organic semiconductor blends. The strongly reduced recombination has been previously attributed to a high probability for the interfacial charge transfer (CT) states (formed upon charge encounter) to dissociate back to free charges. However, whether the reduced recombination is due to a suppressed CT state decay rate or an improved dissociation rate has remained a matter of conjecture. In this chapter an investigation of a donor-acceptor material system that exhibits significantly reduced recombination upon solvent annealing is described. On the basis of detailed balance analysis and the accurate characterization of CT state parameters, it has been shown that an increase in the dissociation rate of CT states upon solvent annealing is responsible for the reduced recombination. This has been attributed to the presence of purer and more percolated domains in the solvent annealed system, which may, therefore, have a stronger entropic driving force for CT dissociation. This chapter is written based on a collaborative work published by the author in the Journal of Physical Chemistry Letters (JPCL) with permission from Ref. [1] copyright(2020) American Chemical Society.

5.1 Introduction

Emerging novel material systems based upon non-fullerene electron acceptors have improved the power conversion efficiency (PCE) of organic solar cells (OSC) up to 18% [2]. However, these efficiencies are limited to laboratory small area devices ($<1\,\mathrm{cm}^2$). The upscaling of OSCs to make industrially viable modules is a challenging task and there are multiple problems to be overcome [3, 4]. One of the problems is related to the junction thicknesses; the photoactive layer of technologically relevant OSCs is usually deposited on a transparent electrode via solution processing of a donor-acceptor blend. While spin coating or similar methods are often used at the lab scale, commercializing large area OSCs requires industry-relevant high throughput deposition methods such as roll to roll processing. Inevitably, the active layer must be sufficiently thick to avoid defects that can partially short-circuit large

© The Author(s), under exclusive license to Springer Nature Switzerland AG 2022
N. Zarrabi, *Optoelectronic Properties of Organic Semiconductors*,
SpringerBriefs in Materials,
https://doi.org/10.1007/978-3-030-93162-9_5

area devices. This is challenging since the performance parameters of organic solar cells, particularly the short-circuit current density (J_{sc}) and fill factor (FF), are highly sensitive to the junction thickness [5–7]. It is well-known that the main reason for the poor performance of thick junction OSCs is low charge collection efficiency due to: (i) a slow charge extraction rate associated with low free carrier mobilities; and (ii) a fast bimolecular recombination rate within the bulk of the active layer. Despite considerable efforts by the community, the mobility of free carriers (electrons and holes) cannot be improved much beyond 10^{-3} cm^{2}V^{-1}s^{-1} in diode architectures relevant to solar cells, because of the disordered nature of organic semiconductors. Yet, a handful of material systems have shown promisingly high efficiencies even when the junction thickness is increased to a micron which is considered to be "very thick" in organic solar cell nomenclature.

As discussed in Chap. 2, recombination of free charges in a material with low mobility can be generally described by the Langevin recombination rate (see Eq. (2.18)) which assumes that the effective recombination rate is determined by the encounter rate of free carries when the Coulomb radius is larger than the hopping distance [8]. It has been shown that the recombination rate in pristine organic semiconductors can be more or less explained by Langevin recombination rate. Conversely, the experimentally determined bimolecular recombination rate constants in donor-acceptor bulk heterojunctions, β_{Bulk}, is generally smaller than the value expected from the Langevin relation (see Eq. (2.20)).

A reduction factor of 0.1 is very common in most BHJ systems [9], while only very few exceptional systems exhibits strongly reduced recombination. One classic example of a material system with suppressed recombination is thermally annealed P3HT:PC$_{60}$BM [10]. This material system delivers a reduced recombination rate of $\sim$100 times less than the Langevin rate ($\gamma = 1/100$), which allows for the FF and the PCE to remain optimal in a range of thicknesses from 80 to 180 nm [11]. In 2014, Sun et al. [12] introduced the donor benzodithiophene terthiophene rhodanine (BTR) which delivered a reduction factor of $\gamma = 1/150$ upon solvent vapour annealing (SVA) when blended with PC$_{71}$BM [13]. This material was arguably the first high efficiency material system exhibiting reduced recombination. It was shown that a 350 nm-thick junction BTR:PC$_{71}$BM solar cell could deliver a high FF ($\sim$0.75) and a PCE of 9.5% as a result of the reduced recombination, even though the electron and hole mobilities remained relatively low and even imbalanced. In 2016, a Naphtho[1,2-c:5,6-c]Bis ([1, 2, 5] Thiadiazole)-based π-conjugated polymer (NT812) was introduced by Jin et al. [14]. Following thermal annealing, a 800 nm-thick junction of NT812:PC$_{71}$BM was found to have a reduction factor of $\gamma = 1/800$, achieving a PCE of >9% [15].

It is important to note that the suppressed recombination in these material systems is only significant if one or more post treatment processes such as thermal or solvent vapour annealing follow the active layer deposition. It has been shown that these treatments influence the active layer morphology, however, a clear structure-property relation is still lacking. In the case of P3HT:PC$_{60}$BM, for example, thermal annealing increases crystallinity and hence the domain sizes in the bulk [16, 17]. As a result, the suppressed recombination rate was initially attributed to the geometrical separation (confinement) of electrons and holes within their respective domains (see Eq. (2.19)).

In general, however, the generation and recombination of electrons and holes in BHJs occurs via charge transfer (CT) states, acting as intermediate charge recombination/generation centres. This correlation between the charge generation quantum yield and the reduction factor has been recently confirmed, experimentally, using a meta-analysis by Shoaee et al. [9]. The authors have shown that the reduction factor depends on the kinetics of the CT states and more specifically on the ratio between dissociation rate constant of the CT states to free carriers k_d as well as the decay rate constant of the CT state to the ground state k_f. However, in systems with strongly reduced recombination, it is not yet known whether the reduction is due to a suppressed k_f or improved k_d. Finding the answer to this question is crucial for a better understanding of non-Langevin recombination and ultimately for obtaining a structure-property relation in this regard-potentially leading to a means to deliver efficient thicker junctions relevant to industrial scaling.

In the work presented in this chapter, the above question has been addressed by introducing a procedure which utilizes the principle of detailed balance and accurate characterisation of CT state parameters. We investigated the donor-acceptor material system BQR:PC$_{70}$BM (benzodithiophene -quaterthiophene-rhodanine: [6, 6]-Phenyl-C$_{71}$-butyric acid methyl ester) which exhibits a strongly reduced recombination of $\gamma = 1/2000$ when solvent annealed. It will be shown that an improved dissociation rate of CT states into free carriers is responsible for the substantial reduction factor in this material system.

5.2 Device Characterisation and Transport Measurements

In Fig. 5.1 the chemical structures of BQR and PC$_{70}$BM are shown.

Two variations of devices were fabricated from the same active layer solution: (1) the as-cast (AC) device for which no post processing was performed on the spin coated active layer; and (2) the solvent vapor annealed (SVA) device for which the spin coated active layer was exposed to THF vapor followed by thermal annealing. Both AC and SVA devices were fabricated with the previously reported structure

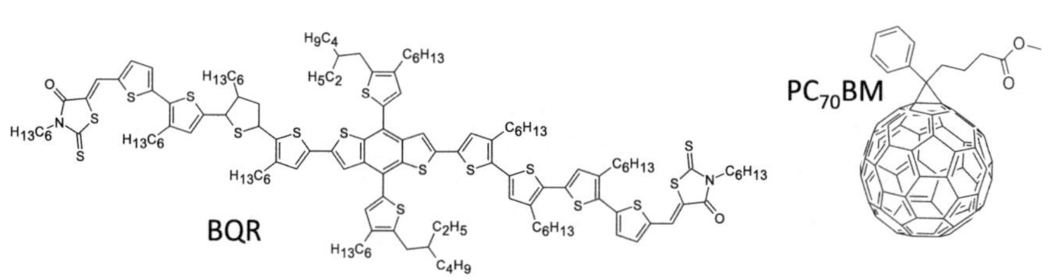

Fig. 5.1 Molecular structure of BQR and PC$_{70}$BM

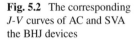

Fig. 5.2 The corresponding *J-V* curves of AC and SVA the BHJ devices

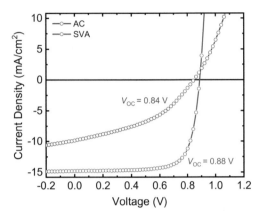

[18] to achieve the optimal device performance (The details of device fabrication can be found in Appendix). The current-voltage characteristics (*J-V*) of 320-nm-thick AC and SVA devices are shown in Fig. 5.2 (see Appendix for the details of the *J-V* measurements). It can be seen that the SVA device has a higher open circuit voltage (V_{OC}), higher short-circuit current (J_{sc}) and higher FF and, thus, higher PCE compare to the AC device. As shown by Schwartz et al., even in a 600-nm-thick junction, the PCE remains higher than 8% when solvent annealed [19].

Investigation of the morphology of the active layers in AC and SVA devices shows that the BHJ structure is sensitive to the active layer deposition process. Grazing-incidence wide-angle scattering (GIWAX) has revealed that solvent vapour annealing increases the crystallinity and the domain purity in the BQR. This results in the formation of small but pure aggregated domains in the blend for both the donor and the acceptor [19, 20]

Transport and recombination measurements were conducted on both AC and SVA devices (see Box 5.1). Both electron and hole mobilities are improved in SVA devices compared to AC which agrees with the purer domains for both donor and acceptor in the SVA blend. The bimolecular rate coefficient for both devices were measured using transient double injection (DI) (see Box 5.2). The β_{Bulk} for AC devices is 10 times reduced ($\gamma = 1/10$), whereas the SVA devices exhibit 2000 times reduced recombination ($\gamma = 1/2000$), compared to the Langevin recombination rate coefficient.

Box 5.1

Resistance Dependent Photovoltage (RPV) Measurement: In order to quantify the charge carrier mobility in the devices we use the resistance dependent photovoltage (RPV) measurement technique [21]. Similar to time of flight (TOF), in this measurement we track the arrival time of photogenerated carriers, known as the transit time, at the electrodes. The schematic of the measurement setup is shown below.

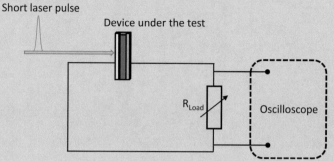

The charge carriers are generated by a short low-intensity laser pulse. We use a Pharos PH1-10 laser at the second harmonic wavelength of 514 nm. The pulse duration was approximately 290 fs and the repetition rate of the experiment was 20 Hz. The fluence of the pump should be low so that the photovolatge at highest load resistance (1 MΩ) remains significantly below the built-in voltage. This ensures that the measurement is performed under short-circuit conditions for all load resistance. The fluence was lowered to less than 1 nJ/cm^2 using neutral density filters. The built-in electric field within the device causes the electrons and holes to drift into their respective electrodes. The measurement is repeated at different load resistances. At low load resistance (low RC constant) the transient of the faster carrier can be observed. The first shoulder that appears in the voltage transient in early times is attributed to the faster carrier. By increasing the load resistance the small contribution of the slower carrier to the current amplifies as a result the voltage transient of the slower carrier will be appear as a second shoulder in later times. The load resistance in our measurement setup was a homemade decade variable-resistance box which can vary the series resistance of the circuit from 50 Ω to 1 MΩ.

We performed RPV measurements for the AC (below figure part a) and SVA (below figure part b) devices. The transit time of the faster and slower carriers are shown for both RPV measurements. In both cases the faster carrier mobility is attributed to the electron mobility. This is consistent with the electron mobility in PC$_{70}$BM that has been reported before which is $\sim 10^{-3}$ cm^{2}V^{-1}s^{-1}. The bars shows the dispersion range for the transit time from which the uncertainty in the values reported for the mobilities can be calculated.

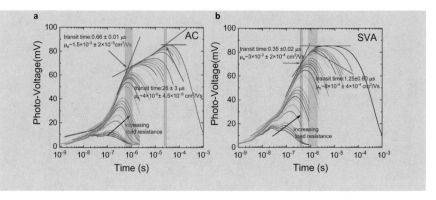

Box 5.2

Transient Double Injection Current Measurement: In order to quantify the bimolecular recombination factor γ in the devices, we utilized the transient double-injection measurements [22]. In this method electrons (from the cathode) and holes (from the anode) are injected into the device. After an RC time which is the time it takes to charge the capacitor electrodes, the total current is given by the double injected J_{DI} current into an insulator (or a semiconductor with low intrinsic charge carrier):

$$J_{DI} = \varepsilon_r \varepsilon_0 \sqrt{\frac{9\pi}{8}} \sqrt{\frac{\mu_p \mu_n (\mu_p + \mu_n)}{\mu_R}} \frac{V^2}{d^3}$$

as shown by Mark and Lampert [23]. Here $\mu_R = \frac{\varepsilon_r \varepsilon_0 \beta_{Bulk}}{2q}$, where β_{Bulk} is the bulk bimolecular recombination rate constant, V is the forward bias voltage and d is the film thickness. If the recombination is governed by the Langevin rate then $\beta = \beta_L$, the electron and hole will immediately recombine (annihilate) following encounters. As such, the total current is the sum of electron and hole space charge limited current:

$$J_{SCLC} = \frac{9\varepsilon_r \varepsilon_0 (\mu_p + \mu_n) V^2}{8d^3}$$

In this case the current does not change with time and it immediately reaches a plateau after the RC time. In contrast, If the recombination is suppressed compared to the Langevin rate then electrons and holes will survive for a longer time as a result the density of carriers (current) will increase with time. Ultimately, the current saturates to a steady state current.

By normalizing J_{DI} to J_{SCLC} concidering (2.18) the reduction factor γ can be calculated as follows:

$$\gamma = \frac{\beta_{\mathrm{Bulk}}}{\beta_{\mathrm{L}}} = \frac{1}{\frac{9(\mu_{\mathrm{p}}+\mu_{\mathrm{n}})^2}{16\pi\mu_{\mathrm{p}}\mu_{\mathrm{n}}} \times (\frac{J_{\mathrm{DI}}}{J_{\mathrm{SCLC}}})^2}.$$

As shown in the following figure γ has been calculated for AC and SVA devices. The saturated $\frac{J_{\mathrm{DI}}}{J_{\mathrm{SCLC}}}$ is also shown in the figure.

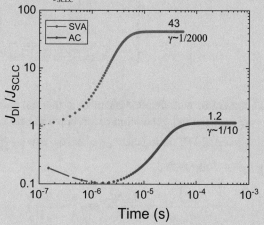

For this measurement a Keysight 33500B function generator was used to apply the voltage pulse to the devices in forward bias and transient double-injection currents were recorded using Rohde&Schwartz RTM3004 oscilloscope.

5.3 Theoretical Framework

The kinetic diagram of a BHJ system is depicted schematically in Fig. 5.3. If we assume that G_{CT} is the generation rate of CT states (either directly or via exciton dissociation), then one can write [24, 25]

$$\frac{dn_{\mathrm{CT}}}{dt} = G_{\mathrm{CT}} - k_{\mathrm{f}}n_{\mathrm{CT}} - k_{\mathrm{d}}n_{\mathrm{CT}} + \beta_{\mathrm{en}}n_{\mathrm{CS}}^2 = 0 \qquad (5.1)$$

under steady-state conditions, where n_{CT} and n_{CS} are the densities of (excited) CT states and separated charge carriers, respectively. The encounter rate, β_{en}, can be slightly reduced relative to β_{L} due to the geometrical confinement of electrons and holes but other morphology-related mechanisms can play roles to reduce it also [26]. Here $k_{\mathrm{d}}n_{\mathrm{CT}} - \beta_{\mathrm{en}}n_{\mathrm{CS}}^2$ is the net generation-recombination rate of free charge carriers

Fig. 5.3 Schematic of charge generation and recombination in a BHJ OSC. The transition illustrated by the dashed downward arrow is the effective recombination rate of separated charge carriers

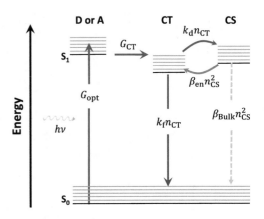

via CT states. In accordance with detailed balance, at thermal equilibrium ($n_{CT} = n_{CT,eq}$ and $n_{CS} = n_{CS,eq}$), the net generation-recombination rate of free separated charge carriers must be zero [27]. Noting that $n_{CT,eq} = N_{CT} \exp\left(-\frac{\epsilon_{CT}}{k_B T}\right)$ and $n_{CS,eq}^2 = N_{CS}^2 \exp\left(-\frac{\epsilon_{SC}}{k_B T}\right)$, it then follows that

$$k_d = \frac{\beta_{en} n_{SC,eq}^2}{n_{CT,eq}} = \frac{\beta_{en} N_{CS}^2}{N_{CT}} \exp\left(-\frac{\Delta\epsilon}{k_B T}\right) \tag{5.2}$$

where $\Delta\epsilon = \epsilon_{CS} - \epsilon_{CT}$ is the energetic offset between CS and CT states; $\epsilon_{CT}(\epsilon_{CS})$ and $N_{CT}(N_{CS})$ are the energy and effective density of states for CT (CS) states, respectively. Furthermore, based on Eq. (5.1) (see Fig. 5.3), the net generation-recombination rate of free charge carriers can be expressed as $k_d n_{CT} - \beta_{en} n_{CS}^2 = G_{CT} P - (1 - P)\beta_{en} n_{CS}^2$, where

$$P = \frac{k_d}{k_d + k_f} \tag{5.3}$$

is the dissociation probability for the CT states. Subsequently, the (effective) recombination coefficient for charge carriers is given by $\beta_{Bulk} = (1 - P)k_f \beta_{en}$ or, equivalently

$$\beta_{Bulk} = \frac{P k_f \beta_{en}}{k_d} \tag{5.4}$$

where k_d is given by Eq. (5.2). In terms of Eq. (2.19), we thus obtain

$$\beta_{Bulk} = \frac{P k_f}{k_d} \times \gamma_{Geo} \tag{5.5}$$

We note that $\gamma_{Geo} \approx 1$, as the geometrical confinement of charge carriers does not play a significant role in the reduction of bimolecular recombination in systems with domain sizes smaller than 10 nm (such as BQR:PC$_{70}$BM [19]) as shown by Heiber et al. [28]. It is thus clear that either increasing k_d or decreasing k_f can result in an improvement in the charge generation efficiency accompanied by a simultaneous reduction of the bimolecular recombination.

The total dark saturation current (of free charge carriers) associated with CT states is equal to the recombination current of free charge carriers at thermal equilibrium and given by: $J_0 = q \int_0^d [(1 - P)\beta_{en} n_{CS,eq}^2] dx = q\beta_{Bulk} n_{CS,eq}^2 d$, where d is the active layer thickness. Using Eqs. (5.2) and (5.4), noting that $J_0 = (J_0^{Rad})/EQE_{LED}$, the associated radiative limit of J_0 is obtained as:

$$J_0^{Rad} = EQE_{LED} \times q d P k_f N_{CT} \exp\left(-\frac{\epsilon_{CT}}{k_B T}\right), \tag{5.6}$$

where EQE_{LED} is the efficiency of the electroluminescence. On the other hand, based on the reciprocity between light absorption and emission, we expect $J_0^{Rad} = q \int_{\epsilon_{min}}^{\infty} EQE_{PV} \times \phi_{BB} d\epsilon$, where EQE_{PV} is the photovoltaic external quantum efficiency, ϵ_{min} is the lower limit of the EQE_{PV} measurement, and ϕ_{BB} is the spectral flux density of the black-body spectrum at room temperature. Furthermore, the radiative limit (V_{OC}^{Rad}) and the non-radiative losses (ΔV_{OC}^{NR}) of the V_{OC} under 1 sun illumination are given by Eqs. (2.28) and (2.31).

Finally, combining Eqs. (5.6) and (2.31), the reduced decay rate constant $P k_f$ is be expressed as:

$$P k_f = \frac{J_0^{Rad}}{q d N_{CT}} \left(\frac{\epsilon_{CT} + q\Delta V_{OC}^{NR}}{k_B T}\right) \tag{5.7}$$

From which the ratio between the reduced k_f s in SVA and in AC devices can be determined via:

$$\frac{P_{SVA} k_{f,SVA}}{P_{AC} k_{f,AC}} = \frac{J_{0,SVA}^{Rad}}{J_{0,AC}^{Rad}} \frac{N_{CT,AC}}{N_{CT,SVA}} \exp\left(\frac{\epsilon_{CT,SVA} - \epsilon_{CT,AC} + q[\Delta V_{OC,SVA}^{NR} - \Delta V_{OC,AC}^{NR}]}{k_B T}\right) \tag{5.8}$$

In the next section using the experimental result we will calculate the ratio and subsequently the $\frac{k_{d,SVA}}{k_{d,AC}}$.

5.4 Experimental Result and Discussion

In Fig. 5.4a the sensitive EQE_{PV} spectra are shown, where $\epsilon_{min} = 1.6\,eV$. The sensitive EQE_{PV} were measured using the method presented in Box 4.1 [29], from which the sub-gap features, in this case the CT sate parameters, can be measured accurately. The CT state absorption for AC and SVA devices are clearly observed at ~1.4 eV. Based on the estimated J_0^{Rad} and J_{Ph}, V_{OC}^{Rad} was calculated using Eq. (2.28) for both AC and SVA devices. These parameters are plotted versus the lower limit of the integration

Fig. 5.4 **a** Sensitive EQE_{PV} is plotted versus photon energy to show the CT state absorption contribution. **b** J_0^{Rad} and V_{OC}^{Rad} and **c** ΔV_{OC}^{NR} for both AC and SVA devices are plotted versus the minimum photon energy at which J_0^{Rad} and hence V_{OC}^{Rad} and ΔV_{OC}^{NR} are calculated

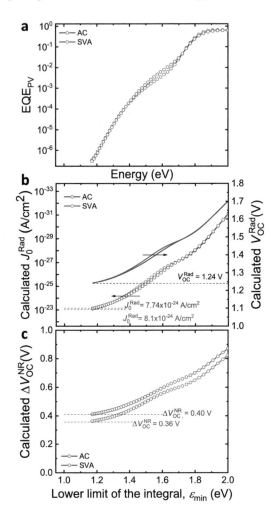

windows, ϵ_{min} (Fig. 5.4b). Furthermore, from the actual cell V_{OC} we determined ΔV_{OC}^{NR} for both devices (Fig. 5.4c).

While it is very difficult to obtain the number density of CT states (N_{CT}), the ratio between N_{CT}s in AC and SVA device can be estimated from the experimental results. To determine the ratio between N_{CT} we first extracted the absorption coefficients from the EQE_{PV}. It has been shown that the Beer-Lambert law cannot be applied for this purpose due to cavity induced interference effects. To eliminate the effect of interference we therefore used the procedure proposed by Kaiser et al. [30] allowing for the sub-gap absorption coefficient spectra to be obtained. This method is based on an iterative inverse transfer matrix formalism for which complex optical constants are required in the visible spectral region. The corresponding refractive indices for AC and SVA devices are provided in Fig. 5.5.

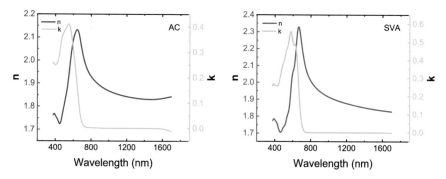

Fig. 5.5 Optical constants: Extinction coefficient **k** and index of refraction **n** of AC and SVA BQR:PC$_{70}$BM films measured via ellipsometry (Box 2.1)

Fig. 5.6 Simulated absorption coefficient for AC and SVA devices plotted versus photon energy. Gaussian fits on the CT state absorption and the CT state parameters obtained from the Gaussian fits (Eq. (4.2)) are shown. Fitting errors are specified for each parameter

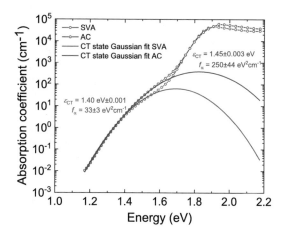

In Figure Fig. 5.6 the absorption coefficients (α_{CT}) are shown together with the Gaussian fits based on Eq. (4.2) and the CT state parameters that have been obtained from the fittings. The uncertainty of ϵ_{CT} are taken from fitting errors and they are specified for each parameter. The pre-factor f_α is obtained from the fittings for each system. Here we assume that f_σ is the same for both AC and SVA devices, made of the same donor and acceptor molecules, thus $\frac{f_{\alpha,AC}}{f_{\alpha,SVA}} = \frac{N_{AC}}{N_{SVA}} = \frac{250}{33} \pm 1.5$ (see supplementry material of Ref. [1]). In this important step we obtained the ratio between the number density of the interfacial charge transfer states, previously extremely challenging to quantify. Our approach is thus far the only methodology to determine the number density of charge transfer states, however, it is only relative and cannot be applied to compare blends of different donor-acceptor molecules with often unknown oscillator strengths.

From the above analysis, the number density of CT states in the AC system is approximately six times larger than that of the SVA. This agrees with the morphological studies on these systems indicating higher domain purity and larger aggregation in the SVA system, and hence smaller number density of CT states in the SVA films.

Table 5.1 Kinetic and thermodynamic parameters of the as-cast and solvent-annealed BQR : $PC_{71}BM$

System	μ_n (cm²V⁻¹s⁻¹)	μ_p (cm²V⁻¹s⁻¹)	$\gamma = \frac{\beta_{Bulk}}{\beta_L}$	V_{OC} (V)	J_0^{Rad} (A/cm²)	V_{OC}^{Rad} (V)	ΔV_{OC}^{NR} (V)	ϵ_{CT} (eV)	ΔV_{OC}^{Rad} (V)
AC	1.5×10^{-3} $\pm 2 \times 10^{-5}$	4×10^{-5} $\pm 4.5 \times 10^{-6}$	0.1 ± 0.01	0.84 ± 0.004	7.74×10^{-24} $\pm 10^{-28}$	1.24 ± 0.00	0.40 ± 0.004	1.45 ± 0.003	0.21 ± 0.003
SVA	3×10^{-3} $\pm 2 \times 10^{-4}$	8×10^{-5} $\pm 4 \times 10^{-4}$	1/2000 ± 0.0001	0.88 ± 0.004	8.10×10^{-24} $\pm 10^{-28}$	1.24 ± 0.00	0.36 ± 0.004	1.40 ± 0.001	0.16 ± 0.001

Table 5.1 summarises all the free charge, CT state and thermodynamic parameters required for calculation of $\frac{P_{SVA}k_{f,SVA}}{P_{AC}k_{f,AC}}$ from Eq. (5.8). From this we find $\frac{P_{SVA}k_{f,SVA}}{P_{AC}k_{f,AC}} = 0.23 \pm 0.07$, meaning that the Pk_f rate of CT states in the SVA system is approximately 4 times slower than that of AC system. This ratio is not small enough to explain the substantial difference of bimolecular recombination in the two systems $\frac{\gamma_{SVA}}{\gamma_{AC}} = 0.005 \pm 0.0002$. Using the rationale $\frac{\gamma_{SVA}}{\gamma_{AC}} = \frac{P_{SVA}k_{f,SVA}}{P_{AC}k_{f,AC}} \times \frac{k_{d,AC}}{k_{d,SVA}}$ as per Eq. (5.5), the ratio of CT states dissociation rates can be found to be $\frac{k_{d,SVA}}{k_{d,AC}} \approx 50 \pm 15$. This implies that the CT states dissociate at a much faster rate in the SVA system than in the AC resulting in significantly reduced recombination. This is the mechanism mainly responsible for the reduced recombination (at least in this material system), while reduction in CT state decay rate has a marginal impact on the reduction factor (roughly a factor of 4). It has been suggested that the dissociation of the CT states at the donor-acceptor interface can be driven by delocalization of charges over a molecule or number of molecules which results in reduced Columbic attraction between electron and hole [31]. As a result, the improvement of CT state dissociation in SVA BQR:PC$_{70}$BM can be attributed to the slightly larger and purer domains of the donor and acceptor which can result in more delocalized charges at the interface.

5.5 Conclusion

In this chapter, a theoretical/experimental methodology to obtain the number density of CT states in one system relative to another was presented. The method is generic, with the limitation that the two systems are required to have similar or known CT states oscillator strengths. The method has been applied on the as cast and solvent annealed BQR:PC$_{70}$BM solar cells with the latter showing significantly reduced recombination rate constants. Using detailed balance analysis together with accurate characterisation of CT state and free charge parameters, we have been able to relate the CT state kinetics to the observed strongly suppressed bimolecular recombination of free charges. The results show that the reduced recombination in SVA devices is due to an improved dissociation rate of CT states upon solvent annealing while a marginal reduction in the CT states decay rate was observed. These results shed considerable light on the nature of reduced bimolecular recombination in BHJ organic solar cells and is a step towards understanding how to engineer the thicker junctions needed for industrially viable OSCs.

References

1. Zarrabi, N., Sandberg, O., Kaiser, C., Subbiah, J., Jones, D., Meredith, P., & Armin, A. (2020). Experimental evidence relating charge-transfer-state kinetics and strongly reduced bimolecular recombination in organic solar cells. *The Journal Of Physical Chemistry Letters, 11,* 10519–10525.
2. Liu, Q., Jiang, Y., Jin, K., Qin, J., Xu, J., Li, W., Xiong, J., Liu, J., Xiao, Z., Sun, K. & Others 18
3. Armin, A., Hambsch, M., Wolfer, P., Jin, H., Li, J., Shi, Z., et al. (2015). Efficient, large area, and thick junction polymer solar cells with balanced mobilities and low defect densities. *Advanced Energy Materials, 5,* 1401221.
4. Meredith, P., & Armin, A. (2018). Scaling of next generation solution processed organic and perovskite solar cells. *Nature Communications, 9,* 5261.
5. Bartesaghi, D., Pérez, I., Kniepert, J., Roland, S., Turbiez, M., Neher, D., & Koster, L. (2015). Competition between recombination and extraction of free charges determines the fill factor of organic solar cells. *Nature Communications, 6,* 7083.
6. Lenes, M., Koster, L., Mihailetchi, V., & Blom, P. (2006). Thickness dependence of the efficiency of polymer: Fullerene bulk heterojunction solar cells. *Applied Physics Letters, 88,* 243502.
7. Min Nam, Y., Huh, J., & Ho Jo, W. (2010). Optimization of thickness and morphology of active layer for high performance of bulk-heterojunction organic solar cells. *Solar Energy Materials And Solar Cells, 94,* 1118–1124.
8. Langevin, P. (1903). Recombinaison et mobilites des ions dans les gaz. *Annales de chimie et de physique, 28,* 122.
9. Shoaee, S., Armin, A., Stolterfoht, M., Hosseini, S., Kurpiers, J., & Neher, D. (2019). Decoding charge recombination through charge generation in organic solar cells. *Solar RRL, 3,* 1900184.
10. Juška, G., Arlauskas, K., Stuchlik, J. & Österbacka, R. (2006). Non-Langevin bimolecular recombination in low-mobility materials. *Journal Of Non-Crystalline Solids, 352,* 1167–1171.
11. Jin, H., Armin, A., Hambsch, M., Lin, Q., Burn, P., & Meredith, P. (2015). Bulk heterojunction thickness uniformity - a limiting factor in large area organic solar cells? *Physica Status Solidi (a), 212,* 2246–2254.
12. Sun, K., Xiao, Z., Lu, S., Zajaczkowski, W., Pisula, W., Hanssen, E., et al. (2015). A molecular nematic liquid crystalline material for high-performance organic photovoltaics. *Nature Communications, 6,* 6013.
13. Armin, A., Subbiah, J., Stolterfoht, M., Shoaee, S., Xiao, Z., Lu, S., et al. (2016). Reduced recombination in high efficiency molecular nematic liquid crystalline: Fullerene solar cells. *Advanced Energy Materials, 6,* 1600939.
14. Jin, Y., Chen, Z., Dong, S., Zheng, N., Ying, L., Jiang, X., et al. (2016). A novel Naphtho[1,2-c?:5,6- c']Bis([1,2,5]Thiadiazole)-based narrow-bandgap π-conjugated polymer with power conversion efficiency over 10%. *Advanced Materials, 28,* 9811–9818.
15. Armin, A., Chen, Z., Jin, Y., Zhang, K., Huang, F., & Shoaee, S. (2018). A shockley-type polymer: Fullerene solar cell. *Advanced Energy Materials, 8,* 1701450.
16. Li, G., Shrotriya, V., Yao, Y., & Yang, Y. (2005). Investigation of annealing effects and film thickness dependence of polymer solar cells based on poly(3-hexylthiophene). *Journal of Applied Physics, 98,* 043704.
17. Sandberg, O., Zeiske, S., Zarrabi, N., Meredith, P., & Armin, A. (2020). Charge carrier transport and generation via trap-mediated optical release in organic semiconductor devices. *Physical Review Letters, 124,* 128001.
18. Geraghty, P., Lee, C., Subbiah, J., Wong, W., Banal, J., Jameel, M., et al. (2016). High performance p-type molecular electron donors for OPV applications via alkylthiophene catenation chromophore extension. *Beilstein Journal Of Organic Chemistry, 12,* 2298–2314.
19. Schwarz, K., Geraghty, P., Mitchell, V., Khan, S., Sandberg, O., Zarrabi, N., et al. (2020). Reduced recombination and capacitor-like charge buildup in an organic heterojunction. *Journal of the American Chemical Society, 142,* 2562–2571.

20. Bourque, A., Engmann, S., Fuster, A., Snyder, C., Richter, L., Geraghty, P., & Jones, D. (2019). Morphology of a thermally stable small molecule OPV blend comprising a liquid crystalline donor and fullerene acceptor. *Journal of Materials Chemistry A, 7,* 16458–16471.
21. Stolterfoht, M., Philippa, B., Armin, A., Pandey, A., White, R., Burn, P., et al. (2014). Advantage of suppressed non-Langevin recombination in low mobility organic solar cells. *Applied Physics Letters, 105,* 013302.
22. Armin, A., Juska, G., Philippa, B., Burn, P., Meredith, P., White, R., & Pivrikas, A. (2013). Doping-induced screening of the built-in-field in organic solar cells: effect on charge transport and recombination. *Advanced Energy Materials, 3,* 321–327.
23. Lampert, M., & Mark, P. (1970). *Current injection in solids.* Cambridge: Academic.
24. Koster, L., Smits, E., Mihailetchi, V., & Blom, P. (2005). Device model for the operation of polymer/fullerene bulk heterojunction solar cells. *Physical Review B, 72,* 085205.
25. Braun, C. (1984). Electric field assisted dissociation of charge transfer states as a mechanism of photocarrier production. *The Journal of Chemical Physics, 80,* 4157–4161.
26. Gorenflot, J., Heiber, M., Baumann, A., Lorrmann, J., Gunz, M., Kämpgen, A., et al. (2014). Nongeminate recombination in neat P3HT and P3HT:PCBM blend films. *Journal of Applied Physics, 115,* 144502.
27. Sandberg, O., & Armin, A. (2019). On the effect of surface recombination in thin film solar cells, light emitting diodes and photodetectors. *Synthetic Metals, 254,* 114–121.
28. Heiber, M., Baumbach, C., Dyakonov, V., & Deibel, C. (2015). Encounter-limited charge-carrier recombination in phase-separated organic semiconductor blends. *Physical Review Letters, 114,* 136602.
29. Zeiske, S., Kaiser, C., Meredith, P., & Armin, A. (2020). Sensitivity of sub-bandgap external quantum efficiency measurements of solar cells under electrical and light bias. *ACS Photonics, 7,* 256–264.
30. Kaiser, C., Zeiske, S., Meredith, P., & Armin, A. (2020). Determining ultralow absorption coefficients of organic semiconductors from the sub-bandgap photovoltaic external quantum efficiency. *Advanced Optical Materials, 8,* 1901542.
31. Bakulin, A., Rao, A., Pavelyev, V., Loosdrecht, P., Pshenichnikov, M., Niedzialek, D., et al. (2012). The role of driving energy and delocalized states for charge separation in organic semiconductors. *Science, 335,* 1340–1344.

Chapter 6
Outlook

Organic semiconductors show considerable promise for photovoltaic energy conversion applications. The photovoltaic power conversion efficiency of the thin film laboratory-scale small-area organic photovoltaics has improved to over 18% in recent years. This significant progress is the result of development in multiple research areas including material design and synthesis, active-layer morphology control, device engineering and, theoretical modeling and understanding that relate the material properties and structure to the device performance. The body of the work presented in this thesis is mostly focused on this last area.

Photogeneration of charge and the photovoltaic effect in organic solar cells are mainly governed by two features of organic semiconductors: Excitonic properties caused by low dielectric constant; and low charge carriers mobility resulting from the disordered nature of organic semiconductors. These two properties complicate the fundamental processes that control current and voltage generation in OPV devices. Understanding the material properties and underlying physics behind basic phenomena in OPV devices can help to further improve the device performance.

Motivated by this, the discussion in this thesis started with a review of the main material properties of organic semiconductors followed by an exploration of the electro-optical phenomena in organic photovoltaic devices. In Chap. 3, a new method for measuring exciton diffusion lengths in organic semiconductors was presented. The method is based on measuring small exciton quenching yields in low-content-quencher (donor or acceptor) devices in the steady-state. The results showed two distinct quenching regimes. At ultra-low quencher densities, an anomalous quenching pathway plays a role which is not efficient but very long range. The long-range quenching can be attributed to the exciton delocalisation right after formation. In the low quencher regime, the results of the modeling and fittings have been used to calculate the exciton diffusion length. The method has been applied to three well-known organic semiconductors being P3HT, PCDTBT, $PC_{70}BM$. The resulting exciton dif-

© The Author(s), under exclusive license to Springer Nature Switzerland AG 2022
N. Zarrabi, *Optoelectronic Properties of Organic Semiconductors*,
SpringerBriefs in Materials,
https://doi.org/10.1007/978-3-030-93162-9_6

fusion lengths are comparable to the values that have been reported in the literature for the material systems under test. The low-impurity content organic solar cells can also be useful for different studies that require minimum morphological alteration to the active-layer. In Chap. 4, these devices were used to investigate and engineer trap states. The result of the study showed that the mid-gap trap states are a universal feature in organic semiconductor donor-acceptor blends. The absorption feature of these states have been directly observed in the EQE_{PV} spectra far below the energy of CT states. Taking these absorption feature into account, a new radiative limit for the open circuit voltage (close to experimentally measured open circuit voltage) has been obtained, which is in apparent violation of reciprocity between photovoltaic external quantum efficiency and electroluminescence quantum efficiency. We propose and provide compelling evidence that mid-gap trap states contribute to photocurrent by a non-linear process of optical release, upconverting them to the CT state. Based on these understandings, a two-diode model was advanced which can accurately describe both the dark current and open circuit voltage. This is a significant extension of reciprocity but reconciles the detailed balance. The results of the study have important consequences for our current understanding of both solar cells and photodiodes. In Chap. 5, using the principle of detailed balance, the kinetic properties of a BHJ material system have been related to the charge transfer state and thermodynamic parameters of the device made from that material system. The outcome was then used to evaluate the ratio between CT state dissociation rate before and after solvent annealing in an exemplary material system $BQR:PC_{70}BM$. This material system exhibits a significantly reduced bimolecular recombination rate relative to Langevin recombination upon solvent annealing. The results of this work provide experimental evidence that the improvement in the dissociation rate of the CT states upon solvent annealing is responsible for the reduced bimolecular recombination rate compared to the Langevin rate.

In summary, the studies presented in this thesis have brought new understanding to some of the fundamental features of organic photovoltaic devices including exciton diffusion and dissociation, charge generation and recombination via mid-gap trap states and, CT state dissociation and free charge carrier dynamics. At the same time, it has proven that there is still a lot more to investigate.

There are a number of major questions arising from the work presented in this thesis that are particularly of interest to the author for future studies:

i: Are excitons more delocalised in new non-fullerene material systems due to crystallinity? In other words are these material systems less disordered?

ii: Is the process of exciton quenching and charge transfer formation different in the new low offset donor-acceptor material systems? In other words are charge transfer states mediated in charge generation?

iii: What is the effect of mid-gap trap state in thick junction photovoltaic devices?

iv: Are there any relevances between material properties like disorder and non-raditive losses in organic solar cells?

v: Given the fact that recombination current via mid-gap trap states is dominant at low light intensities, are organic solar cells proper candidate for indoor photovoltaic applications?

vi: Can new organic material systems deliver high efficiency in large scale thick devices?

The answer to these questions and many others could be new steps towards full understating of organic semiconductor devices which can lead to an improvement in the device performance, and maybe even a viable and important new clean energy technology.

Appendix
Extra Experimental Methods

A.1 Device Fabrication

A.1.1 Materials

PEDOT:PSS (Poly(3, 4-ethylenedioxythiophene)-poly(styrenesulfonate)) was purchased from Heraeus.

Zinc acetate dehydrate was purchased from Sigma Aldrich.

PCDTBT (Poly[N-9"-heptadecanyl-2, 7-carbazole-alt-5, 5-(4', 7' -di-2-thienyl-2', 1', 3'-benzothiadiazole)]) was purchased from Sigma Aldrich.

PCPDTBT (Poly[2, 6-(4, 4-bis-(2-ethylhexyl)-4H-cyclopenta[2, 1-b; 3, 4-b')-dithiophene)-alt-4, 7 -(2, 1, 3-benzothiadiazole)]) purchased from Sigma Aldrich.

O-IDTBR was purchased from Sigma Aldrich.

PC$_{70}$BM ([6, 6]-Phenyl-C71-butyric acid methyl ester) was purchased from Solarmer (Beijing).

PDINO (perylene diimide functionalized with amino N-oxide) was purchased from Solarmer (Beijing).

EH-IDTBR was purchased from Solarmer (Beijing).

BQR (benzodithiophene-quaterthiophene-rhodanine) was provided by Dr. David. J Jones (University of Melbourne).

m-MTDATA (4, 4', 4'-Tris[(3-methylphenyl) phenylamino]triphenylamine) was purchased from Ossila.

PM6 (Poly[(2, 6-(4, 8-bis(5-(2-ethylhexyl-3-fluoro)thiophen-2-yl)- benzo[1, 2-b:4, 5-b']dithiophene))-alt-(5, 5-(1', 3'-di-2-thienyl-5', 7'-bis(2-ethylhexyl)benzo[1', 2'-c:4', 5'-c']dithiophene-4, 8-dione)]) was purchased from Zhi-yan (Nanjing) Inc.

Y6 (2, 2'-((2Z, 2'Z)-((12, 13-bis(2-ethylhexyl)-3, 9-diundecyl-12, 13-dihydro-[1, 2, 5] thiadiazolo[3, 4-e]thieno[2", 3':4', 5']thieno[2', 3':4, 5]pyrrolo[3, 2-g]thieno[2', 3':4, 5]thieno[3, 2-b]indole-2, 10-diyl)bis(methanylylidene))bis(5, 6-difluoro-3-oxo-

N. Zarrabi, *Optoelectronic Properties of Organic Semiconductors*,
SpringerBriefs in Materials,
https://doi.org/10.1007/978-3-030-93162-9

2, 3-dihydro-1H-indene-2, 1-diylidene))dimalononitrile) was purchased from Zhi-yan (Nanjing) Inc.

ITIC (3, 9-bis(2-methylene-(3-(1, 1-dicyanomethylene)-indanone))-5, 5, 11, 11-tetrakis(4-hexylphenyl)-dithieno[2, 3-d:2', 3'-d']-s-indaceno[1, 2-b:5, 6-b'] dithiophene) was purchased from Zhi-yan (Nanjing) Inc.

PBDB-T (Poly[(2, 6-(4, 8-bis(5-(2-ethylhexyl)thiophen-2-yl)-benzo[1, 2-b:4, 5-b']dithiophene))-alt-(5, 5-(1', 3'-di-2-thienyl-5', 7'-bis(2-ethylhexyl)benzo[1', 2'-c:4', 5'-c']dithiophene-4, 8-dione)]) was purchased from Zhi-yan (Nanjing) Inc.

PTB7-Th (Poly[4, 8-bis(5-(2-ethylhexyl)thiophen-2-yl)benzo[1, 2-b; 4, 5-b'] dithiophene-2, 6-diyl-alt-(4-(2-ethylhexyl)-3-fluorothieno[3, 4-b]thiophene-)-2-carboxylate-2-6-diyl)]) was purchased from Zhi-yan (Nanjing) Inc.

TAPC (4, 4'-Cyclohexylidenebis[N, N-bis(4-methylphenyl)benzenamine]) was purchased from American Dye Source. **P3HT** (Poly(3-hexylthiophene-2, 5-diyl)) was purchased from Merck.

A.1.2 Substrate Preparation

Commercial patterned indium tin oxide (ITO) coated glass substrates from Ossila were used for all devices. All the substrates were cleaned in an Alconox (detergent) aqueous solution bath at 60 °C, followed by sequential sonication in deionize (DI) water, acetone and 2-propanol for 10 min each. The cleaned substrates were dried with nitrogen and then treated in UV-Ozone cleaner (Ossila, L2002A2-UK).

A.1.3 Electron/Hole Transport Layer (ETL/HTL) Deposition

Solar cells were fabricated with either a conventional or inverted architecture. For the conventional devices, PEDOT:PSS was used as the HTL. A PEDOT:PSS solution was first filtered through a 0.45 μm PVDF filter, then it was spin-coated (6000 rpm for 30 s resulting in a thickness of 30 nm) onto ITO substrates and annealed at 155 °C for 15 min. For the inverted devices, zinc oxide (ZnO) was used as the ETL. A ZnO solution was prepared by dissolving 200 mg of zinc acetate dihydrate in 2-methoxyethanol (2 ml) and ethanolamine (56 μl). The solution was stirred overnight under ambient conditions and was spin-coated onto ITO substrates (4000 rpm resulting in a thickness of approximately 30 nm). The substrates were annealed at 200 °C for 60 min.

A.1.4 Active Layer and Top Electrode Deposition

The deposition methods of the active layers are described below for each sample. All top electrodes were deposited by thermal evaporation under a vacuum of 10^{-6} Tor with an appropriate mask (from Ossila) to define a 0.04 cm^2 cell area for each Pixel.

TAPC:PC$_{70}$BM low-donor-content: devices were fabricated with a conventional architecture (ITO/PEDOT:PSS/TAPC:PC$_{70}$BM/Al). A solution of PC$_{70}$BM in chloroform with the concentration of 32 mg mL^{-1} was initially prepared and stirred at 30 °C for one hour and filtered afterwards. A solution of TAPC in chloroform was also prepared with the highest blend ratio needed (1% wt). This solution was then diluted sequentially to obtain all the lower blend ratios. The PC$_{70}$BM and TAPC solutions then combined in 1:1 v/v ratio to give the blend solution with the concentration of 16 mL^{-1}. The final solutions were spin coated on the pre-prepared substrate at 2500 RPM for 25 s to achieve 60 nm thickness of the active layer.

PCDTBT:PC$_{70}$BM low-acceptor-content: devices were fabricated with a conventional architecture (ITO/PEDOT:PSS/PCDTBT:PC$_{70}$BM/Al). A solution of PCDTBT in chlorobenzene with the concentration of 20 mg mL^{-1} was initially prepared and stirred at 100 °C for one hour. A solution of PC$_{70}$BM in chlorobenzene was also prepared with the highest blend ratio needed (20% wt). This solution was then diluted sequentially to obtain all the lower blend ratios. The PCDTBT and PC$_{70}$BM solutions were then combined in 1:1 v/v ratio to give the blend solution with the concentration of 10 mg mL^{-1}. The final solutions were heated to 100 °C and cooled down to ambient temperature and finally spin coated on the pre-prepared substrate at 600 RPM for 60 s to obtain 60 nm thickness of the active layer.

P3HT:PC$_{70}$BM low-acceptor-content: devices were fabricated with a conventional architecture (ITO/PEDOT:PSS/P3HT:PC$_{70}$BM/Al). A solution of P3HT (21 kDa, RR = 93%, PDI = 1.7) in chloroform with the concentration of 20 mg mL^{-1} was initially prepared and stirred at 40 °C for one hour and filtered afterwards. A solution of PC$_{70}$BM in chloroform was also prepared with the highest blend ratio needed (0.5% wt) and stirred at 30 °C for one hour and subsequently filtered. This solution was then diluted sequentially to obtain all the lower blend ratios. The P3HT and PC$_{70}$BM solutions were then combined in 1:1 v/v ratio to give the blend solution with the concentration of 10 mL^{-1}. The final solutions were spin coated on the pre-prepared substrate at 2500 RPM for 30 s to achieve 60 nm thickness of the active layer. P3HT can form a crystalline or partially crystalline phase in the active layer. It has been shown that fullerene molecules can be repelled from the crystalline phase which makes the assumption of bulk quenching invalid. We should note that in order to minimize the crystallization we used low-molecular-weight P3HT (21 kDa) in our work and deposited it via a fast-drying process using chloroform as the solvent.

BQR:PC$_{70}$BM devices were fabricated with a conventional architecture (ITO/ PEDOT: PSS/BQR:PC$_{70}$BM/Ca/Al). For as cast devices, BQR and PC$_{70}$BM were dissolved in toluene (24 mg/ml with the donor:acceptor ratio of 1:1) and stirred at 60 °C for 3 h. Then BQR:PC$_{70}$BM solution was spin coated (1000 rpm) on the

PEDOT:PSS layer to achieve a film thickness of 100 nm. For solvent annealed (SVA) devices, the BQR:PC$_{70}$BM films were further exposed to a Tetrahydrofuran (THF) environment in a closed petri dish for 20s and then thermally annealed (90 °C) for 10 mins. For both SVA and as cast devices, 20 nm of calcium (Ca) and 100 nm of Aluminium (Al) were evaporated as the top electrodes.

PCDTBT:PC$_{70}$BM:m-MTDATA devices were fabricated with an inverted architecture (ITO/ZnO/PCDTBT:PC$_{70}$BM:m-MTDATA/MoO3/Ag). 30 mg of PCDTBT: PC$_{70}$BM with a blend ratio of 1:4 (i.e. 6 mg of PCDTBT and 24 mg of PC$_{70}$BM) was firstly dissolved in 800 µl of Chlorobenzene (CB) (3 batches). 200 µl of a solution containing 0.06 mg, 0.006 mg, and 0 mg of m-MTDATA (Mw = 789.02 g/mol) was then added to the first solutions in order to have the final solutions containing 1%, 0.1% and 0% by weight of m-MTDATA in PCDTBT. The solution was spin-coated using a spin rate of 800 rpm to obtain an active layer thickness of 90 nm. 7 nm of MoO3 and 100 nm of Ag were then evaporated as the top electrode.

PCDTBT:PC$_{70}$BM devices were fabricated with a conventional architecture (ITO/ PEDOT:PSS/PCDTBT: PC$_{70}$BM/PDINO/Ag). PCDTBT and PC$_{70}$BM were dissolved in Dichlorobenzene (DCB) with the donor:acceptor ratio of 1:4, and the thicknesses of the active layers were adjusted by changing the concentration of the solution and the speed of spin-coating (30 mg mL^{-1} DCB solution with 1500 rpm for 54 nm active layer, 40 mg mL^{-1} DCB solution with 1500 rpm for 85 nm active layer, 40 mg mL^{-1} DCB solution with 1000 rpm for 105 nm active layer, 40 mg mL^{-1}1 DCB solution with 600 rpm for 155 nm active layer, 50 mg mL^{-1} DCB solution with 600 rpm for 185 nm active layer, 60 mg mL^{-1} DCB solution with 600 rpm for 315 nm active layer, 60 mg ml-1 DCB solution with 400 rpm for 585 nm active layer). 10 nm of PDINO was cast on the active layer from a methanol solution (1 mg mL^{-1}), then 100 nm of Ag was deposited on the PDINO to form a cathode.

PM6:Y6 devices were fabricated with an inverted architecture (ITO/ZnO/PM6:Y6/ MoO3/Ag). PM6:Y6 was dissolved in a CF solution (14 mg ml-1 with 0.5 vol.% CN) with a donor:acceptor ratio of 1:1.2, and spin-coated (3000 rpm) on ZnO to form 100 nm film. The cast active layers were further treated with thermal annealing at 110 °C for 10 min. 7 nm of MoO3 and 100 nm of Ag were evaporated as the top electrode.

PM6:ITIC devices were fabricated with an inverted architecture (ITO/ZnO/PM6: ITIC/MoO3/Ag). PM6:ITIC was dissolved in a CB solution (18 mg $^{-1}$ with 0.5 vol.% DIO) with a donor:acceptor ratio of 1:1, and spin-coated (1000 rpm) on ZnO to form 100 nm film. The active layers were further treated with thermal annealing at 100 °C for 10 min. 7 nm of MoO3 and 100 nm of Ag were evaporated as the top electrode.

PM6:O-IDTBR devices were fabricated with an inverted architecture (ITO/ZnO/ PM6: O-IDTBR/ MoO3/Ag). PM6:O-IDTBR was dissolved in a CB solution (18 mg ml^{-1}) with a donor:acceptor ratio of 1:1, and spin-coated (1000 rpm) on ZnO to form 100 nm film. 7 nm of MoO3 and 100 nm of Ag were evaporated as the top electrode.

PBDB-T:EH-IDTBR devices were fabricated with an inverted architecture (ITO/ ZnO/ PBDB-T: EH-IDTBR/ MoO3/Ag). PBDB-T:EH-IDTBR was dissolved in a CB solution (14 mg ml^{-1}) with a donor:acceptor ratio of 1:1, and spin-coated (800 rpm) on ZnO to form a 100 nm film. 7 nm of MoO3 and 100 nm of Ag were evaporated as the top electrode.

PBDB-T:ITIC devices were fabricated with an inverted architecture (ITO/ZnO/ PBDB-T:ITIC/MoO3/Ag). PBDB-T: ITIC was dissolved in a CB solution (14 mg ml^{-1} with 0.5 vol.% DIO) with a donor:acceptor ratio of 1:1, and spin-coated (800 rpm) on ZnO to form 100 nm film. The active layers were further treated with thermal annealing at 100 °C for 10 min. 7 nm of MoO3 and 100 nm of Ag were evaporated as the top electrode.

PTB7-Th:ITIC devices were fabricated with an inverted architecture (ITO/ZnO/ PTB7-Th: ITIC/MoO3/Ag). PTB7-Th:ITIC was dissolved in a CB solution (14 mg ml^{-1} with 1 vol.% DIO) with a donor:acceptor ratio of 1:1.4, and spin-coated (1000 rpm) on ZnO to form 100 nm film. 7 nm of MoO3 and 100 nm of Ag were evaporated as the top electrode.

PBDB-T:PC$_{70}$BM devices were fabricated with an inverted architecture (ITO/ZnO/ PBDB-T: PC$_{70}$BMM/MoO3/Ag). PBDB-T:PC$_{70}$BM was dissolved in a CB solution (14 mg ml^{-1} with 3 vol.% DIO) with a donor:acceptor ratio of 1:1.4, and spin-coated (1000 rpm) on ZnO to form 100 nm film. Then the as-cast films were rinsed with 80 μL of methanol at 4000 rpm for 20 s to remove the residual DIO. 7 nm of MoO3 and 100 nm of Ag were evaporated as the top electrode.

PTB7-Th:PC$_{70}$BM devices were fabricated with an inverted architecture (ITO/ZnO/ PTB7-Th:PC$_{70}$BM/ MoO3/Ag). PTB7-Th:PC$_{70}$BM was dissolved in a CB solution (14 mg ml^{-1} with 3 vol.% DIO) with a donor:acceptor ratio of 1:1.5, and spin-coated (600 rpm) on ZnO to form 100 nm film. Then the as-cast films were rinsed with 80 μL of methanol at 4000 rpm for 20 s to remove the residual DIO. 7 nm of MoO3 and 100 nm of Ag were evaporated as the top electrode.

PCPDTBT:PC$_{70}$BM: PCPDTBT:PC$_{70}$BM devices were fabricated with an inverted architecture (ITO/ZnO/PCDTBT:PPC$_{70}$BM/MoO3/Ag). PCPDTBT:PC$_{70}$BM was dissolved in a DCB solution (40 mg ml^{-1}) with a donor:acceptor ratio of 1:4, and spin-coated (1500 rpm) on ZnO to form 80 nm film. 7 nm of MoO3 and 100 nm of Ag were evaporated as the top electrode.

Crystalline silicon solar cell: Commercial crystalline silicon solar cell (Part number: KXOB22-12X1).

Germanium Photodiode: Purchased from Newport (818-IR).

A.2   EQE$_{PV}$ Measurement

EQE spectra of the devices were recorded using a PV Measurements Inc. QEX7 set-up with monocromator light power on the order of $\sim \mu$W cm^{-2}, which was calibrated by a NREL-certified photodiode without light bias.

A.3   Dark *J-V* Measurement

A Keithley source-measure unit (model 2400) with a home-built software was used to accurately (very sensitive to low current) measure the dark current-voltage characteristics of the samples. The scan speed used for this measurement was 0.1 pps.

A.4   White Light *J-V* Measurement

The light J–V characteristics were obtained using a Keithley 2400 source and measurement unit with a home-built software under 1 sun illumination (AM 1.5G, $\sim$100 mW cm^{-2}). A Newport solar simulator (M94011A) calibrated with a reference cell and meter (91150V from Newport) was used as the light source. The scan speed used for this measurement was 0.1 pps.

A.5   Photoluminescence Measurement

Photoluminescence measurements were conducted using the fundamental (1030 nm) of Pharos PH1-10W laser as a pump (laser power 40 mW cm^{-2}). The photoluminescence spectrum of the sample was measured using a photonic multi-channel analyser (PMA) from Hamamatsu (model C10028) with corresponding software provided by the company (U6039-01 version 4.1.2).

A.6   Intensity Dependent Open-Circuit Voltage and Photocurrent Measurement

Intensity dependent photocurrent measurements were performed using a 4-channel fiber-coupled laser source (Thorlabs, MCLS1-CUSTOM) with variable output power. The excitation wavelength was set to 1550 nm and no bias voltage was applied on the device (short-circuit). A Keithley 2450 was used to record the light intensity dependent device photocurrent, while the incident light power was recorded by a NIST-calibrated photodiode sensor (Newport, 818-IR). Photocurrent density versus open-circuit voltage (V_{OC}) measurements, on the other hand, were performed at an excitation wavelength of 520 nm (using a commercial laser) in combination with a Keithley 2450 used to record both photocurrent (short-circuit) and open-circuit voltage of the device. The incident light intensity was varied by using a motorized attenuator (Standa, 10MCWA168-1) containing different optical density filters.

A.7 EQE$_{LED}$ Measurement

EQE$_{LED}$ of the solar cell devices were measured using a Hamamatsu EQE measurement system C9920-12. An integrating sphere was used as the sample chamber in order to account for different radiation angle and absorption of the sample. A Keithley source-measure unit (model 2400) was used to drive the electroluminescence of the samples. Depending on the wavelength range of the EL of samples, two different spectrometers (from 346 to 1100 nm and from 896 to 1688 nm spectral range) were used to detect the electroluminescence. The software (U6039-06 Version 4.0.1) for the EQE$_{LED}$ measurement and calculation was provided by the Hamamatsu.

Printed in the United States
by Baker & Taylor Publisher Services